D1575724

"'50 Ways to Improve Product Reliability' is a great collection of ground-rules that are based upon experience. It describes a number of situations and presents practical examples of what engineers should do to be successful in the area of reliability. In a few examples it also indicates what should not be done in order to avoid problems. The book is organized according to the phases of a project and a maturity matrix is presented as a means to measure progress. Many examples show proactive use of reliability tools for hardware and software. All are accompanied by concise case studies showing how to apply tools and handle common situations. It is a must read for all engineers and managers who are involved with corporate-wide reliability improvement efforts."
James McLinn, CRE, Fellow, ASQ

"All engineering products have to be designed and manufactured with reliability targets having been taken into account during the product development process. From an engineering perspective, the role of reliability within product design and development is one that applies a systematic process of proven engineering techniques throughout the design program. In modern engineering programs, it is no longer sufficient to view reliability engineering as simply a verification activity; reliability cannot be tested into any product, it has to be designed in. Therefore, the role of reliability engineering within a product design environment can be defined as controlling warranty costs by assisting design engineers to make great products better. A key component within any effective product development process is Design for Reliability, which when deployed enables reliability principles to be employed to evaluate the inherent reliability of the product being designed; it also identifies potential opportunities for reliability improvement. In this book, Mike Silverman delivers valuable guidance on design, analysis, and product improvement activities that can be used by engineers to optimize the reliability of the products they are designing, and ultimately to insure customer satisfaction. This book is an invaluable resource for anyone that is involved with engineering great products, from corporate managers to reliability and design engineers alike. Mike Silverman takes you on a guided tour of Design for Reliability techniques, which are interwoven with many fascinating and useful case studies that demonstrate how this engineering approach has already benefited many organizations. In providing you with '50 Ways to Improve Product Reliability,' Mike has gifted you with the tools to help make your great products better."
Mark Turner, C.R.P.,
Six Sigma & DFSS Black Belt,
Manager of Reliability,
Enecsys

"After reading Mike's '50 Ways to Improve Product Reliability,' I am impressed to state that if most people who want reliable products would execute fifty or sixty percent of what he is espousing, how great it would be for consumers. Instead of going into depth on each topic, Mike stays at a higher level to give a perspective of reliability within the product development lifecycle. If you need more in-depth knowledge on any of the techniques discussed in his book, you can easily contact a consulting firm such as Mike's company, Ops A La Carte®, and get the educational or consulting expertise you need. I have known Mike for over 15 years, and he is one of the most honest people I know. Feel free to get to know him and you will see what I am talking about. Enjoy his book!"

Harry McLean
Manager of Reliability, Advanced Energy Inc.
Author of *HALT, HASS, and HASA Explained: Accelerated Reliability Techniques*

How Reliable is Your Product?

50 Ways to Improve Product Reliability

By Mike Silverman
Foreword by Patrick O'Connor

E-mail: info@superstarpress.com
20660 Stevens Creek Blvd., Suite 210
Cupertino, CA 95014

Published by Super Star Press™, a Happy About® imprint
20660 Stevens Creek Blvd., Suite 210, Cupertino, CA 95014
http://superstarpress.com/

First Printing: December 2010
Hardcover ISBN: 978-1-60773-060-6 (1-60773-060-X)
eBook ISBN: 978-1-60773-061-3 (1-60773-061-8)
Place of Publication: Silicon Valley, California, USA
Paperback Library of Congress Number: 2010941885

Trademarks

Warning and Disclaimer

Acknowledgments

I would like to thank my staff of consultants who have contributed immensely to the content of this book. I would like especially to acknowledge the following individuals for contributing or helping to review the book: Lennox Bennett, Marc Bush, Doug Farel, George de la Fuente, Craig Hillman, Ted Kalal, Andre Kleyner, Lou LaVallee, Bob MacLevey, Harry McLean, Jim McLeish, Jim McLinn, Bob Mueller, Pat O'Connor, Kim Parnell, Fred Schenkelberg, Bernard Silverman, Bryan Stallard, Mark Turner, and Arthur Zingher. I would also like to thank the late Dr. Gregg Hobbs for his pioneering efforts in the area of HALT and HASS and Harry McLean for his innovative work in the area of HASA and in his development of the HALT-to-AFR Calculator.

I would also like to thank my family for their patience and support during the development of this book.

Contents

Figures

Foreword by Patrick O'Connor

Modern engineering products, from individual components to large systems, must be designed, developed, and manufactured to be reliable in use. Designs must be robust in relation to the stresses and other factors that could cause damage or deterioration in transport, storage, use, and maintenance. Product development must include testing to ensure that this is achieved and to show up weaknesses for correction. The manufacturing processes must be performed correctly and with the minimum of variation. All of these aspects impact the costs of design, development, manufacture, and use, or, as they are often called, the product's life cycle costs. The challenge of modern competitive engineering is to ensure that life cycle costs are minimized while achieving requirements for performance and time to market. If the market for the product is competitive, improved reliability can generate very strong competitive advantages, as well as cost savings to manufacturers and to users. Today, this message is well understood by most engineering companies that face competitive pressures.

The customers for major systems, particularly the U.S. military, drove the quality and reliability methods that were developed in the West from the 1950s onwards. They reacted to perceived low achievement by the imposition of standards and procedures. The methods included formal systems for quality and reliability management (MIL-Q-9858 and MIL-STD-758) and methods for predicting and measuring reliability (MIL-STD-721, MIL-HDBK-217, and MIL-STD-781). MIL-Q-9858 was the model for the international standard on quality systems (ISO9000). The methods for quantifying reliability were similarly developed and applied to other types of products and have been incorporated into other standards such as ISO60300. The application of these approaches has been controversial and not always effective.

In contrast, the Japanese quality movement that began in the 1950s was led by an industry that learned how manufacturing quality provided the key to greatly increased productivity and competitiveness, principally in commercial and consumer markets. The methods that they applied were based upon understanding of the causes of variation and failures, as well as continuous improvements through the application of process controls and motivation and management of people at work. It is one of history's ironies that the foremost teachers of these ideas were Americans, notably

P. Drucker, W.A. Shewhart, W.E. Deming, and J.R Juran. The Japanese also applied methods for design for reliability, notably Quality Function Deployment (QFD) and failure modes and effects analysis (FMEA).

By the turn of the century, methods of design for reliability and for manufacturing quality excellence had become refined. Most of the U.S. military standards were discontinued. More practical and effective methods were applied almost universally, particularly by industries whose products faced international competition or other drivers, particularly high costs of failures or strict customer requirements. However, some still cling to unrealistic mathematical precision for predicting and measuring reliability, as well as to bureaucratic approaches to quality management.

In the same time frame, there have been improvements in design capabilities with advances in computer-aided engineering, as well as in materials and in manufacturing processes. We have seen dramatic improvements in the reliability of products as diverse as automobiles, telecommunications, domestic equipment, and spacecraft. How many readers have experienced a failure of a microprocessor or an automobile engine?

I am pleased to endorse and recommend this new book. Mike Silverman presents a wealth of practical, experienced-based wisdom in a way that is easy to read and apply. He has avoided detailed descriptions of methods, emphasizing instead the management and team aspects of applying cost-effective reliability improvement tools in ways that work.

The main methods he covers include reliability planning, design techniques (FMEA, fault tree analysis), test—particularly highly accelerated life test (HALT), and design of experiments, as well as methods for reliability prediction, stress derating, vendor reliability, failure reporting and analysis, and others. The whole product life cycle is considered, from initial design through prototype test, manufacturing, and field service, to obsolescence. He emphasizes the need for integration of reliability efforts to ensure their effective application. The fifty chapters all include brief case histories that illustrate this.

I recommend the book as an excellent guide for engineering project management and their teams, as well as for reliability specialists. It demystifies the sometimes difficult methods and helps specialists to communicate with managers, designers, and other engineers. It will make your products more reliable.

Patrick O'Connor
November 2010

Why Am I Writing This Book?

I've read many reliability and quality textbooks, and very few approach reliability from the practical perspective. Instead, these books are filled with theory and formulas. However, many engineers are starting with almost no knowledge on the subject of reliability; they are in need of some basic education, but even more, they need the benefit of some practical experience and guidance. I wrote this book as a helpful guide, and I targeted the book at engineering professionals around the world in need of a practical guide to reliability.

I wrote this book based on my 25 years' practicing reliability, including 10 years running a reliability test lab and 10 years running a reliability consulting firm called Ops A La Carte®. I started Ops A La Carte® because I saw the need to teach and help companies develop reliability programs. Most engineers I come across know basic concepts and have their favorite reliability techniques, but few understand how to put this into an overall reliability program.

Ops A La Carte® has worked with over 500 different companies in over 100 different industries in 30 different countries, so we have the ability to provide guidance from the experiential point of view. When I use the collective term "we" in this book, I am referring to an experience we have had at Ops A La Carte®.

Just like any other discipline, there is no substitute for experience. Book knowledge is a good start, but until you are working on a design program or faced with a particular failure situation, you may not know what to do, or you may panic and resort to ineffective techniques you used in the past. In this book, I will show you different techniques and give you real-life situations that we faced and how we used particular techniques to solve problems.

I saw a movie recently called *Eagle Eye* that is quite applicable to reliability. The movie starts with the discovery of a possible terrorist plot. The Joint Chiefs of Staff of the United States consult their new supercomputer "Eagle Eye" to determine the probability that the terrorist plot is real. The supercomputer comes back with a probability of 39%. The commander in charge responded by saying, "39% and probability don't belong in the same sentence." (So true…and very appropriate for reliability as well.) In the next scene, the Joint Chiefs consult with the President, and by this time, "Eagle

Eye" collects a bit more information and raises its probability to 51%. The President then decides to take action based on this and authorizes an attack on the alleged terrorist group. What is that really telling us? In fact, a probability of 51% means that there is a 49% chance that the conclusion is incorrect based on the data.

Likewise, with reliability tests, you need to make decisions based on test data from a sample of the population. You will never have enough data to be 100% certain of any decision, so you should gain as much confidence as you can with the time and money that you have. That is the art of reliability testing.

I structured the book in 50 easy-to-read chapters. Each chapter has some background on the reliability technique, its usefulness, and in some cases, its limitations. In addition, when applicable, I compare the technique in question to other techniques to show you when to use which technique. Starting in Chapter 3, I introduce the topic of Reliability Integration, and for each chapter onwards, I comment on how you can use the concept of Reliability Integration with that particular technique. I will talk a lot about Reliability Integration. It is one of the most valuable takeaways from this book.

In each chapter, I will provide one or more case studies from clients we have worked with and discuss how we utilized the specific technique in question. I didn't use the names of people or companies, but all of the case studies are real.

Tips on how to best use this book:

- If a phrase is highlighted in bold italics, that means the term is a main technique of that chapter and is in the table of contents as well as the index. I will also capitalize the phrase throughout the rest of the book as an indication that it is an important technique. For all other important terms, check the index for other places I have used the same term.
- I included a guide to acronyms. The field of reliability uses a lot of acronyms and I know how frustrating it can be reading a book filled with them.
- I included a glossary of terms.

If you feel I missed something or you have information to add to a particular topic, I'd love to hear from you. I hope you enjoy my book.

Part I
Introduction

RELIABILITY INTEGRATION^{SM}

CONCEPT
DESIGN
PROTOTYPE
MANUFACTURING

Reliability Engineering Services Integrated Throughout the Product Life Cycle

1 Guidelines, Not Rules

Reliability is an interesting discipline because there are many techniques you can use to solve problems and create a reliable product. There certainly are guidelines and best practices, but you should determine for yourself the set of techniques that will work within your company. Some of the factors you should consider are the size of your company, company culture, past experiences, background education, marketplace, and customer requirements.

Even though this book is filled with different techniques, don't think you need to use them all. Everything I write in the rest of the book is in the form of guidelines and tips. You should determine for yourself which of these tips will work for you and which won't when developing your reliability program. Also, don't try to copy someone else's reliability program, even if it is from the same industry, and don't copy a reliability program from a previous company. They can be great starting places and can be very valuable input, but trying to copy one program will only get you into trouble because what makes a reliability program work has as much to do with the people involved and the culture of the organization as it does with the product that you are working on. I've seen two companies making almost identical products have completely different reliability programs, and both programs worked for the respective companies.

The important thing to remember is that whatever reliability program you put together should have metrics so you can measure where you are at any given time in the product development process, and the program should produce positive results.

CASE STUDY: Guidelines, Not Rules

I was performing a Reliability Assessment for a military subcontractor, and as part of the assessment, I asked to see their Reliability Program Plan (RPP). Our client was so proud of the plan because of its sheer volume—it was over 100 pages! I asked the reliability engineer who wrote the plan how it came to be. He said that he picked up one of the more well-known reliability textbooks, opened up the table of contents, and made each chapter of the book a section of his plan. I looked at his plan more closely and noticed that he copied the entire plan almost verbatim. He thought that the closer he followed the book, the more successful he would be. This couldn't be further from the truth. The plan should have a good foundation, and a reliability textbook is as good as any other source, but then the plan should be tailored to fit your particular needs. This plan obviously wasn't. Fortunately, his customer wasn't easily fooled and rejected the plan.

We then worked with our client to tailor his plan to match his organization's particular situation. His customer accepted this new plan. Our client followed the plan and developed a very reliable product.

2 What is Design for Reliability (DFR)?

Reliability is no longer a separate activity performed by a distinct group within your organization. Product reliability goals, concerns, and activities are integrated into nearly every function and process of your organization. Senior management's role is to foster an environment where your team keeps reliability and quality goals clearly in mind. Engineering teams should balance project costs, customer maintenance costs, quality, schedule, performance and reliability (and possibly other factors specific to your industry) to achieve optimal product designs. Your organization's structure should encourage all members of your team to apply appropriate reliability methods and principles. The Design for Reliability (DFR) role for the reliability personnel is often finding the cost effective components and design structures with minimal risk and then presenting this to the rest of your team.

When I got started in the reliability field in 1984, reliability seemed to be this "throw it over the wall" concept in which design engineers passed the completed design over to our reliability group. By the time we got the product, we could do little to affect the design, so the reliability effort was mostly focused on measuring where we were. If the product didn't meet its goals at that point, what could we do? We would ship the product and then spend the next several years monitoring its performance and fixing what we could. Gradually, the product got better over time. Our company used our customers as the feedback mechanism. The end result was we had unhappy customers and we had a reputation for poor reliability. Unfortunately, this practice used to be very common.

In today's global economy, so many industries are competing on reliability; we have realized that this "throw it over the wall" method can't work anymore. Reliability should be designed in. What better way to design in reliability than to make the designers responsible for the reliability of their designs? This concept has caught on so well that today, when I lecture on DFR, my audience is largely made up of designers, not reliability engineers. The role of the reliability engineer is changing into the mentor. The reliability engineer is now responsible for going out and finding the best techniques to use and then training the designers on how to use them. The reliability engineer is responsible for writing the Reliability Program Plan (RPP), and the designers are responsible for executing the plan.

DFR is about getting the designers to take ownership of the reliability of the product. The reliability department then becomes the steering committee, helping to set the policies, providing the direction, and training. I liken it to a rowing race. The reliability engineer is the coxswain. (On a rowing team, the coxswain is the member who sits in the stern facing the bow, steers the boat, and coordinates the power and rhythm of the rowers.) The designers are the rowers. When the two work together well, the boat seems to glide across the water smoothly and effortlessly.

2.1 Who Should Take a DFR Course?

Every time I teach a DFR course, my clients always ask me who should attend. Well, the designers of course. Should marketing attend? Sure, they should. You should have representation from sales, customer service, and manufacturing as well. Of course, the reliability team should also attend. Anyone who has a stake in the reliability of the product should attend. However, the designers are a key to the process. Each day-to-day decision they make about the design of the product will ultimately affect the reliability of the product. What is the role of the reliability group? That group should be the leaders and the educators, the ones that help create the goals and write the RPP. Therefore, the day-to-day activities should be performed by the designers (being guided by the reliability team).

CASE STUDY: Getting Your Designers to Buy Into DFR?
A telecommunications company was suffering from low product reliability. Each time they discovered a problem either in product testing or in manufacturing, the test engineers and manufacturing engineers couldn't get the designers to help fix the problem because the designers were too busy designing the next product. You see, they were being incentivized for how fast they could bring the product to market, not for how reliable the product was. Then the company did a smart thing—the CEO made a change and told the designers that they are now responsible for helping fix field failures. You know what happened? All of a sudden, the designers started listening to the reliability team, and a DFR program was born inside their company. Most designers love to work on new product designs, and most dislike having to redesign something that didn't work quite right. The sooner they can get on to the next project, the better from their perspective. Therefore, tying their next project to the successful reliable completion of the previous project is a good incentive for them to ensure they design a reliable product.

---

3 Reliability Integration Provides Integrity

Reliability Integration is the process of seamlessly and cohesively integrating reliability techniques together to maximize reliability at the lowest possible cost. What this means is you should think of your reliability program as a set of techniques that are used together rather than just a bunch of individual activities.

You are building a system, and a system is made up of different components and assemblies; there are different disciplines involved (some of the main disciplines are electrical, mechanical, software, firmware, optical, and chemical). All of the individual pieces make up the system, so don't forget about the interactions, and make sure that you think of the reliability from a system perspective. In Figure 3.1, we illustrate this point using the disciplines of electrical, mechanical, and software.

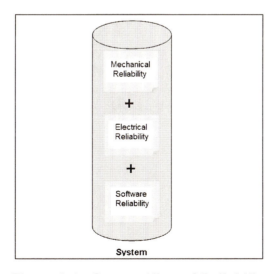

Figure 3.1: System View of Reliability

This is especially true of software versus hardware disciplines. Most companies work on Software Reliability and Hardware Reliability separately and don't integrate the two. When failures occur, this then results in finger-pointing rather than synergy.

This is equally true of electrical versus mechanical disciplines. We see more synergy between these two groups during programs than between software and hardware; however, at the beginning, they rarely get together to discuss common reliability goals and how to apportion them down to each major area of the system.

Product development teams view reliability within each of the separate sub-domains of mechanical, electrical, and software issues. Your customers view reliability as a system-level issue, with minimal concern placed on the distinction between mechanical, electrical, and software issues. Your customer wants the whole product and all its parts to work together perfectly. Since the primary measure of reliability is made by your customer and their end users, engineering teams should maintain a balance of both views (system and sub-domain) in order to develop a reliable product.

3.1 Reliability versus Cost

Intuitively, the emphasis in reliability to achieve a reduction in warranty and in-service costs results in some minimal increase in development and manufacturing costs. Use of the proper techniques during the proper life cycle phase will help to minimize total life cycle cost (LCC).

To minimize total LCC, your organization should do two things:

1. Choose the *best techniques* from all of the techniques available, and apply these techniques at the proper phases of the product life cycle.
2. Properly *integrate* these techniques by feeding information between different phases of the product life cycle.

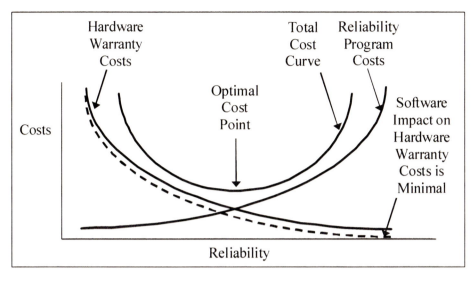

Figure 3.2: System Reliability versus Cost

In Figure 3.2, it is evident that:

1. Program costs go up as you spend more on reliability. At a certain point, you won't get your return on investment (ROI) because the reliability has reached a point where it is becoming increasingly more difficult to improve. That is why it is important to know what the goal is, and it can be just as detrimental to your company to produce a product that is *too* reliable as *not reliable enough*. The product that is too reliable usually comes with increased costs; your customers may not need this level of reliability and will opt for the less expensive product. When was the last time you purchased a $200 blender or toaster?

2. Warranty costs go up as reliability goes down.

3. Software has no associated manufacturing costs (other than perhaps the cost of CDs and manuals and the cost of personnel to test the product in production), so warranty costs and savings are almost entirely allocated to hardware. If there is no cost savings associated with improving Software Reliability, why not leave it as is and focus on improving hardware reliability to save money? You shouldn't do this for two reasons:

   a. Our experience is that for typical systems, software failures outnumber hardware failures by a ratio of 10:1 (see Section 31.1 for more details). Customers buy integrated systems, not just hardware.

   b. The benefits for a Software Reliability program aren't in direct cost savings. Instead, the benefits are in:

i. Increased software/firmware staff availability with reduced operational schedules, resulting in fewer corrective maintenance events.

ii. Increased customer goodwill based on improved customer satisfaction.

CASE STUDY: Linking Electrical, Mechanical, and Software Reliability together

We were working with a semiconductor equipment company to help improve their reliability on their next generation product. First, we provided a Design for Reliability (DFR) seminar for each of the three different disciplines—the electrical group, the mechanical group, and the software group. Then, we met with the electrical, mechanical, and software team leads and developed reliability goals. We started with high level system goals and the apportioned the goals down to each subsystem—electrical, mechanical, and software. Each group lead then took the goal for his subsystem and broke it down further within his area. We worked with each group lead to put together a reliability program plan to meet his subsystem goals. We rolled each of these different subsystem plans into an overall reliability plan for the product. We worked with each group lead to ensure he was on track for meeting his subsystem goals throughout the product development process. The end result was that our client was able to achieve reliability goals for each subsystem and for the system as a whole.

4

Balance between In-House and Outside Help for Your Program

When developing a reliability program, you should decide if you are going to staff up, if you are going to seek outside help, or both. Most companies choose both. They use in-house staff for daily tasks, such as going to meetings and working side-by-side with the design team, and they use outside expertise in areas where they either don't have the expertise or sufficient staff to execute the tasks. Here are the top ten reasons (in no particular order of importance) why you may want to bring in a reliability consultant to give your company a boost:

1. Reliability Leader
2. Reliability Assessment
3. Technology Risk Assessment
4. Specialized Skill-Set
5. Training for Your Engineers
6. During Ramp-Up
7. After a Layoff
8. Facilitation
9. Geographic Help
10. Expert Opinion

4.1 Reliability Leader

You need a strong reliability leader to set a direction for your reliability program, develop corporate metrics, and determine what types of people you need to put in place to make your program successful. In Chapter 2, we discussed the topic of Design for Reliability (DFR), but that will only work if the designers feel they are being led by a competent reliability leader. If your company lacks this reliability leadership, it is time to bring in an expert consultant to lead your team and to provide training. If you already have reliability engineers in-house, you may want to employ a "train-the-trainer" methodology so that the consultant trains your reliability team and then one of the members of the team steps up and assumes the role of the leader to lead your organization through the program.

4.2 Reliability Assessment

Often times, you look back at the past few product releases and come to the conclusion that your product reliability needs improvement. It is never one specific issue, yet the little issues add up to customer dissatisfaction. You need a more predictable approach to reliability. In this case, an outside firm can come in and assess where you are weak. We call this a Reliability Assessment. We talk more about this technique in Chapter 5.

4.3 Technology Risk Assessment

If you are entering a new market or using a new type of technology, you need to ensure that you will be able to achieve a specific level of reliability. A good example of this is deciding between solid state memory and traditional hard drives. What is the inherent reliability of this technology? Is it better or worse than traditional hard drives? You will need an expert to come in and perform a Technology Risk Assessment to evaluate the environment in which you are deploying your product to determine the expected reliability.

4.4 Specialized Skill-Set

If your customer requests a specific technique (or if internally you determine you need to use a specific technique) and you don't know how to do it, then it is time to call in an expert. Using the technique Design of Experiments (DOE) is a good example when you may want to call in an expert. Most companies don't perform DOE often enough to become proficient at it, and therefore your team may be a bit rusty on how to perform one effectively. An outside company can provide a "hands on" training course in which they teach you and provide a workshop to help you implement one at the same time. The reliability field is fairly vast and very few people are versed in all aspects of this field. It is very common to have situations occur that are outside of the ability of well-rounded engineers. When this occurs, it is time to call in a specialist.

4.5 Training for Your Engineers

Experts can come in and not only provide a training course but also lay out a training curriculum for your company. This curriculum can fill in the training gaps for all your engineers and help your newly hired engineers get up to speed quickly. This curriculum should also contain an element of "train-the-trainer" so that you start developing the expertise in-house to provide future training.

4.6 During Ramp-Up

When a company finds itself needing to ramp up quickly in the product development area and either can't hire fast enough or isn't sure if the extra personnel is needed short-term or long-term, then it is best to bring in consultants to ease the load. It is much easier to bring in a consultant than to hire an employee (often, you can bring in a consultant within a few days of when you identify the need and that individual should be able to hit the ground running). It is also very easy to let a consultant go (in California, for instance, it is quite difficult to lay off a regular employee and the risk of a lawsuit can be very high, but with a consultant, this risk is reduced greatly).

4.7 After a Layoff

After a layoff, you can effectively use outside consultants because you will still be developing new products and the work still needs to get done. This helps to reduce your total cost because you are not carrying a full-time employee but instead only paying for consultants when you need them.

4.8 Facilitation

Frequently, you have technical skills in-house but lack facilitation skill. A good example of this is with the technique of Failure Modes and Effects Analysis (FMEA). Most engineers are familiar with this very basic technique of figuring out how a product may fail and then determining ways to mitigate the potential failures. The problem is that each engineer is taught a slightly different technique, so when you get a group of engineers in a room with various methodologies, it can get quite chaotic. What you need is a strong leader to guide you through the process and make sure that everyone's ideas are heard and no one individual dominates the conversation. We discuss facilitation more in Chapter 17.

4.9 Geographic Help

There are occasions in which you need expertise in a specific geographical region because of the location of your company or your vendors. It may make sense to hire local talent rather than flying one of your engineers halfway around the world for two reasons.

A. it is much less expensive to use local talent, not only because of the savings in travel, but often because your other division or your vendors are in an area with lower salary ranges, allowing you to take advantage of less expensive local talent; and

B. local talent is fluent in the language and understands the local customs. Some of the larger reliability consulting firms have a worldwide presence and may have the local talent needed in that part of the world.

4.10 Expert Opinion

If your customers need an expert second opinion, hiring a consulting firm is an effective way to accomplish this. In most cases, company engineers are on the right track and their methodology or plan or report just needs a bit of tweaking. We have even seen situations in which a consultant comes in and says the same thing as an in-house engineer, but just because s/he is a consultant, people within the company or your customer may put more value on the consultant's statement.

RELIABILITY INTEGRATION: Integrating Outside Help with Your In-House Engineering Staff
One logistical concern some companies have when outsourcing is how to outsource a particular activity without requiring a lot of in-house engineering to manage the external resources. For this reason, it is important at the start of the project to outline the relationship (including the roles and responsibilities of the in-house and external resources), how the reviews will take place, and how often. If done properly, external resources can be a huge asset to your team. If not, managing the external team can be a drain on your in-house resources.

CASE STUDY: Expert Reliability Review Service Targeted at the Asia Market
The consulting market in Asia is quite a bit different than that of the United States. We started consulting in Singapore in 2006 and learned that Singapore and Asia as a whole don't hire consultants quite the same way that U.S. companies hire consultants. Companies in Asia tend to want to keep the work in-house and typically only hire consultants for training or when they need specialized skills. Rarely do they ever hire for reasons of lack of bandwidth. Currently there is no lack of engineers in the Asian workforce. For this reason, we developed an Expert Reliability Review service, and we targeted this service at the Asian market. What this entails is the company engineers perform the work, then we review their work, make any changes necessary, and sign off on the engineers' work. This gives our client's customer a level of confidence that they wouldn't have previously had, and the review is quick and quite affordable.

Part II
Marketing/Concept Phase

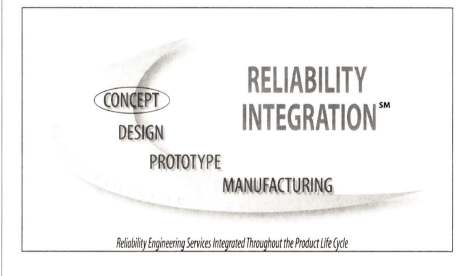

CONCEPT
DESIGN
PROTOTYPE
MANUFACTURING

RELIABILITY INTEGRATION[SM]

Reliability Engineering Services Integrated Throughout the Product Life Cycle

5 Assess Your Assets

Before you can recommend a reliability program for a product, you should first determine the capability of your current organization. We call this a **Reliability Program Assessment**.

5.1 What is a Reliability Program Assessment?

A Reliability Program Assessment is a detailed evaluation of your entire organization's approach and processes across all departments that are involved in creating your products. The assessment captures the current state of your organization and leads to an actionable Reliability Program Plan (RPP).

A Reliability Program Assessment doesn't have to be performed before the start of every program, but it should be performed for any of the following reasons:

1. An established company is trying aggressively to improve their reliability due to a large number of failures in-house and/or in the field.
2. An established company is spending too much money on warranty returns and needs to find a way to reduce it.
3. An established company doesn't know its own stage of reliability on its products or doesn't know why its products are being returned.
4. An established company is trying to get into a new market.
5. A new company is developing its first product.
6. A company, established or new, has never written a reliability plan.

5.2 Motivation for Performing an Assessment

A Reliability Program Assessment identifies systemic changes that impact reliability. It ties into the culture of your organization and to the product. It also provides a roadmap for activities that achieve results. It is the matching of capabilities and expectations.

A good analogy is going to the doctor when you are sick. How can a doctor prescribe medicine to make you better unless he/she first assesses the situation? The same is true for a reliability program. Unless you understand what is wrong and why, you won't know how to fix the situation. When you visit a doctor, the medication that s/he prescribes will depend not only on your symptoms but also on you, the patient. Some patients react better to one type of medication than another. The same is true for a reliability program. Some company cultures will accept certain types of techniques or certain types of changes, and others may not. It is up to the assessor to determine what will work in that particular culture. For example, some companies are very strong in the area of statistics and can use statistics very well in solving problems. For these types of cultures, adding techniques such as using Weibull Analysis for problem solving to determine time to failure will fit into the culture quite easily. However, if the engineers don't have this background, then introducing statistical techniques such as this may fail the adoption process.

5.3 Steps in a Reliability Program Assessment

A Reliability Program Assessment consists of the following steps:

1. Select People to Survey
2. Select the Survey Topics
3. Develop a Scoring System
4. Analyze Data
5. Review Results with Participants (Check-Step)
6. Summarize Results
7. Recommend Actions
8. Assess Particular Areas in More Detail

5.3.1 Select People to Survey

The first step is to determine who the appropriate people to survey are. This is largely based on the goals of your assessment. The survey is meant to be an inquiry, or a fact-finding discovery process. The job of the assessor is to obtain as much information from the individual as possible. The following are typical people we recommend including for the assessment process:

- Design engineers of all disciplines (e.g., electrical, mechanical, optical, chemical, software, firmware)
- Engineering managers
- System engineers
- Manufacturing engineers
- Test engineers (hardware and software)
- Quality and reliability engineers
- Vice president of corporate quality
- Procurement
- Program managers
- Sales/marketing
- Product support/field service
- Key suppliers
- Contract Manufacturer (CM) or Original Design Manufacturer (ODM)

Note that this list may vary depending on roles and responsibilities within your organization.

5.3.2 Select the Survey Topics

You should have at least 20–30 questions covering different aspects of reliability within your organization, including Management, Design, and Manufacturing. The following are examples of the types of questions to ask:

- Under the topic of "Management," one question could be "To what extent are reliability goals (such as annualized failure rate (AFR) or warranty rate) established?"
- Under the topic of "Design," one question could be "To what extent is Failure Modes and Effects Analysis (FMEA) used?"
- Under the topic of "Manufacturing," one question could be "To what extent is Design for Manufacturability (DFM) employed?"

Note that it is important to get responses to the same question from several different people to get a good cross-functional response from the team. It is possible that one person you are talking with has a distorted view of what is going on. For example, if you ask management about Goal Setting, they may respond that it is performed at the start of every project, and the requirements are developed by marketing and passed onto engineering. Then if you ask engineering, you may find that this process isn't happening at all. You may even need to dismiss some survey responses if it is obvious the person you are surveying is upset at the company and uses this survey process as a means for expressing their frustration.

5.3.3 Develop a Scoring System

You should develop a scoring system for the assessment. To do this, you need to come up with a score for each response. The following is a suggested scoring system.

Scoring: 5 = extremely effective
4 = effective
3 = moderately effective
2 = slightly effective
1 = not effective
0 = not done or discontinued
- = not visible, no comment

The score relates to the effectiveness of the technique you are discussing. The effectiveness is a measure of how well the technique is used and how often. If the team performs the technique well when they use it, but they use it only sporadically, it is still not an effective technique. Note that if the person being surveyed has no visibility in the process, you should allow them to pass and give them a "no score" rather than a zero. This then isn't included in the overall score. In order to compile results, combine more than one response for each question and reach an overall score.

During one of our assessments, we asked the team the question "How Important is the FMEA technique?" The manufacturing engineer's response was that it is only as a troubleshooting technique, and for this, we gave it a score of 2. The design engineer's response was that it is only important when evaluating critical design elements, and for this, we gave it a score of 3. The reliability engineer's response was that it is an important design technique and is used on all portions of the design, and for this, we gave it a score of 4.

We then averaged each of these three scores for a combined total of 3.0. However, more than the score, you should see a pattern of an issue here. If the reliability engineer thinks this is an important design technique, then why doesn't the design engineer use it more often? Surely he/she should be part of the FMEA process if it is such a critical design technique. Perhaps it isn't as effective as the reliability engineer thinks it is, and if it is an important technique in the development process, then there is a big hole in the process that you need to address.

5.3.4 Analyze Data

After the completion of an assessment, you should review the data and average the scores to determine the results. This can be as simple as averaging the scores for each question or you may decide to add a weighting factor to emphasize specific areas. Once you have analyzed each of the responses, you should then categorize each response and summarize them.

5.3.5 Review Results with Participants (Check-Step)

The next step is a very important one—reviewing the results with the team and providing an opportunity for additional responses. This is what we call the check-step. It is also known as the "Did we get that right?" step. The purpose of the assessment is to capture the results as accurately as possible, and if you miss something during the surveys, you need to make sure you acknowledge that and give those you surveyed a chance to make additions. Remember that if you don't perform the assessment process effectively and don't capture the true behavior of your organization accurately, you don't stand a chance of being able to set the direction for the organization.

5.3.6 Summarize Results

Next, summarize all of the comments for each question, and provide feedback to the individuals involved in the surveys. You need to capture each of the results given, along with the resulting score. In Section 5.3.6.1, we describe a summary tool we developed called a Reliability Maturity Matrix.

5.3.6.1 Reliability Maturity Matrix

The Reliability Maturity Matrix is a method of categorizing responses and coming up with a summary of where your organization is compared to the rest of the industry. Tables 5.1a and 5.1b show a table we call the Reliability Maturity Matrix process. We created our matrix based on the theme of Crosby's quality maturity methodology from the 1970s. You are welcome to use this matrix, but you may want to modify the matrix depending on your industry.

Measurement Category	Stage I: Uncertainty	Stage II: Awakening	Stage III: Enlightenment	Stage IV: Wisdom	Stage V: Certainty
Management Understanding and Attitude	Management has no comprehension of reliability as a management technique. Management blames reliability engineering for "reliability problems"	Management recognizes that reliability may be valuable, but they aren't willing to provide money or time to make it happen.	Management is still learning more about reliability. They are becoming supportive and helpful.	Management participates and understands absolutes of reliability. They recognize their personal role in continuing emphasis.	Management considers reliability an essential part of company system.
Reliability within the Organizational Chart	Reliability consists of a single engineer who may be doubling as a quality engineer or another function. Reliability is hidden in manufacturing or engineering departments.	A stronger reliability leader is appointed, yet the reliability function is still buried within manufacturing or engineering departments.	Reliability manager reports to top management with a role in management of division.	Reliability manager is an officer of the company and is involved with consumer affairs.	Reliability manager is on the board of directors. Reliability is a thought leader.
Problem handling	Organization is in firefighting mode; no root cause analysis or resolution takes place.	Teams are set up to solve major problems. Long-range solutions aren't identified or implemented.	Corrective action process is in place. Problems are recognized and solved in an orderly way.	Problems are identified early in their development. All functions are open to suggestions and improvements.	Except in the most unusual cases, problems are prevented.

Measurement Category	Stage I: Uncertainty	Stage II: Awakening	Stage III: Enlightenment	Stage IV: Wisdom	Stage V: Certainty
Cost of Reliability as % of net revenue	Warranty: unknown Reported: unknown Actual: 20%	Warranty: 4% Reported: unknown Actual: 18%	Warranty: 3% Reported: 8% Actual: 12%	Warranty: 2% Reported: 6.5% Actual: 8%	Warranty: 1.5% Reported: 3% Actual: 3%
Feedback process	No reliability testing is performed. No field failure reporting other than customer complaints and returns.	Some understanding of field failures and complaints. Designer engineers and manufacturing don't get meaningful information.	Reliability manager reports to top management with a role in management of division.	Refinement of testing systems—only testing critical or uncertain areas. Increased understanding of causes of failure allows deterministic failure rate prediction models.	The few field failures are fully analyzed, and product designs or procurement specifications are altered. Reliability testing is performed to augment reliability models.
DFR program status	No organized activities occur. Organization has no understanding of such activities.	Organization is told reliability is important. DFR techniques and processes are inconsistently applied and only occur when time permits.	Implementation of DFR program with thorough understanding and establishment of each technique.	DFR program is active in all areas of company—not just design & manufacturing. DFR is a normal part of research and development and manufacturing.	Reliability improvement is a normal and continued activity.
Summation of reliability posture (actual quotes from companies)	"We don't know why we have problems with reliability."	"Is it absolutely necessary to always have problems with reliability?"	"Through commitment and reliability improvement, we are identifying and resolving our problems."	"Failure prevention is a routine part of our operation."	"We know why we don't have problems with reliability."

We have divided the Reliability Maturity Matrix into five different stages and we have created seven different categories from which to determine these stages of maturity. Read across each row and find the statement that seems most true for your organization. The average of the stages is your organization's overall maturity stage. Quite often, the level of maturity will vary depending on the category, but it is unusual for an organization to vary by more than one or two levels from one measurement category to the next.

It is rare to find companies that are strictly Stage 1. Often when we approach companies with the idea of performing an assessment, their response is, "Why bother, we already know we do absolutely nothing in the way of reliability." Most companies have some form of reliability program, but it usually isn't documented and isn't well practiced. Companies usually have pockets of excellence within their organization that we can extract and highlight.

It is also rare to find companies that are strictly Stage 5. Stage 5 is reserved for "best in class" companies, and there aren't too many of these that exist (and for the ones that do exist, they probably won't share their techniques with the rest of the world because this is a competitive advantage). Stage 5 companies have their processes down so well that they don't need assessments to tell them where they are and typically don't call in outside consulting teams to assess their processes—they already know where they are.

5.3.7 Recommend Actions

After you get the feedback and feel confident that you have captured the essence of your organization, it is time to come up with recommendations. This is perhaps the most difficult portion of the assessment. This is where experience really counts. You should have a good understanding of "best practices" in the industry to be able to discover a pattern and draw a conclusion from the pattern. If you have never performed an assessment before, I recommend that you call in an expert for your first assessment to make sure you perform it thoroughly and to make sure you draw the proper conclusions. From the assessment, you are looking for trends, gaps in processes, skill mismatches, over-analysis, and under-analysis. Look for differences across your organization, pockets of excellence, areas with good results, and areas that need work.

No one technique or set of techniques makes an entire reliability program. The techniques need to match the needs of your products and your culture.

Many companies that we score a Stage 2 or Stage 3 ask us what they need to do to reach Stage 5. First of all, achieving Stage 5 is quite rare. Secondly, moving more than one stage within one product release is also rare. You should set your expectations appropriately and be patient while you change your systems to achieve better reliability. If you try to make changes too quickly, it is likely that the changes will be rejected by your team or your reliability program will start breaking down.

5.3.8 Assess Particular Areas in More Detail

During the assessment, there are bound to be areas that require further investigation. A list of survey topics may not be enough to assess the situation and provide recommended feedback. For these areas, we recommend a more detailed assessment. Some of these areas may be:

1. Review of field results
2. Review of manufacturing yields
3. Assessment of key supplier(s)
4. Review of design control process

You can then add the results from these detailed assessments to your overall recommended actions.

5.4 Observations from over 100 Assessments

We have performed a Reliability Assessment for over 100 different companies or organizations. Based on our observations, no two companies are alike. Even two different divisions within the same company could be very different. Therefore, when performing an assessment for your company, we advise you to assess each division separately and not just assume that one division is representative of the entire company. Based on all our assessments, we determined the best and worst areas for companies. They are listed in the next two sections. I have highlighted many of these later in the book either as a separate chapter or as important topics within a chapter. Refer to the table of contents and the index for more information on a particular topic of interest in Sections 5.4.1 and 5.4.2.

5.4.1 What Companies Are Doing Best

1. Reliability Predictions
2. HALT
3. Fast reaction to fix problems
4. Golden Nuggets
5. Software bug tracking database
6. Software system testing

5.4.2 What Companies Are Weak At/Opportunities for Improvement

1. Goal Setting
2. Reliability planning
3. Repair & warranty information is invisible
4. Lessons Learned capture process
5. Single owner of product reliability
6. Multiple defect tracking systems
7. Good DFR techniques
8. Use of statistics
9. Synergy between hardware and software engineering

RELIABILITY INTEGRATION: Integrating Goals with a Reliability Program

The Reliability Program Assessment drives the rest of the activities in your reliability program. The recommendations from the assessment help guide your company's future action for your entire reliability program. At suitable points during your program, it also makes sense to reassess your situation to determine how much the gap has shrunk and how close you are to achieving your goals.

CASE STUDY: Assessment Warned Client to Stay Away from Market

Our client made test equipment for the engineering test lab environment, and they were contemplating entering a new market using similar equipment for control systems. The reliability requirements of lab equipment are significantly lower than that of control systems. Using the assessment process, we showed them that they weren't ready to undertake this new type of product. It was the equivalent of jumping two to three stages of reliability maturity in one product release; this typically requires a complete organization restructure (and they weren't prepared to perform this restructuring). Based on our recommendation, they decided not to enter this new market. Overall, they saved a great deal of money and protected their reputation.

6 Goals Aren't Just for Soccer and Hockey

Goal Setting is the process of setting targeted goals at the beginning of a design/development program and then putting forth a plan to achieve the goals. Of all the companies we have worked with, only a handful have done this well. Most companies don't even bother with this process, and the ones that do attempt Goal Setting usually go about it all wrong. They think of Goal Setting as a nice to achieve target and go in with the attitude, "I will try to meet my goals." In the words of Yoda in the movie *Star Wars*: "Try not. Do, or do not. There is no try."

For many companies, this is where you begin with reliability in a product development life cycle. The assessment is a great place to tell you where you are *before* you begin a new project, but the Goal Setting is really the first reliability technique you will implement *during* a new project.

What does a goal statement look like? A reliability goal includes each of the four elements of the reliability definition:

1. Probability of product performance
2. Intended function
3. Specified life
4. Specified operating conditions

Example of a solid reliability goal: *A desktop business computer in an office environment with 95% reliability for one year.*

Let's break down this statement into the four elements:

1. Probability of product performance: *95%*
2. Intended function: *Desktop business computer (and the functions that go along with this type of application)*

3. Specified life: *One year*
4. Specified operating conditions: *Office environment*

Reliability goals can be derived from the following methods:

1. Customer-Specified Goals
2. Internally-Specified Goals
3. Benchmarking Against Competition

6.1 Customer-Specified Goals

For Customer-Specified Goals, your customers will specify the reliability re-quirements for your product. Mean time between failure (MTBF), mean time to repair (MTTR), Availability, dead on arrival (DOA) rate, and return rate are common reliability goals, but there are many others.

Sometimes, your customers won't specify the exact requirements, but rather they will imply requirements such as:

1. "The product should be '*highly reliable*' over its life."
2. "The product shouldn't fail in a way that requires a mission to be aborted."
3. "No loss of data is allowed."

Listening to your customers is perhaps the easiest way to set goals. However, I caution you first to assess your situation before you blindly write down a goal that you may have no way of achieving.

6.2 Internally-Specified Goals

Internally-Specified Goals are usually based on trying to be better than previous products. One of the executives of a company may put forth some sort of edict such as "our next product will have half the field returns than our previous product." You can then turn this into a goal statement.

Often, goals coming from within a company are a bit more subtle. If no one has made a bold statement as in the previous paragraph, you are on your own to determine what you believe you can achieve. To develop a goal, you should conduct Goal Setting meetings with key members of various departments and at your contract manufacturing (CM) partner in order to find out what they have set as internal goals and what they believe they can achieve on your next product. You should then combine these into overall system goals.

You may need to adjust these goals as you gather information, but this work represents a good starting point. New goals from customers may supersede any internal goals. Information from the Gap Analysis may cause you to change your goals. If the gap is unrealistically high, it may make sense to reduce goals so that they are attainable. Often, internal improvement goals require changes to development processes. Goals less than two times better than your current performance can usually be achieved by adjustments to existing processes. Goals more than two times better than the current performance usually require significant changes to existing processes or the adoption of new development practices.

6.3 Benchmarking Against Competition

Your competitors are often the best source of goals, especially if your competitors are the current leaders in the market. Benchmarking Against Competition is a way to determine if your goals are similar to that of your competitors. If not, make adjustments as needed.

Which Goal Setting method is right for your product? I recommend that you use all three—Customer-Specified Goals, Internally-Specified Goals, and Benchmarking Against Competition. Don't just use goals provided to you by your customers. Use all methods to create the most complete and accurate goals possible.

RELIABILITY INTEGRATION: Integrating a Lessons Learned Process with Your Goal Setting Process
Quite often we see companies develop a reliability program and implement all of the techniques internally, then hire an external consultant to tie the techniques together and facilitate the Lessons Learned process. Tying the techniques together helps to ensure that your reliability program is cohesive and integrated throughout the organization.

CASE STUDY: Goal Setting Starts at the Beginning of the Program

In a project with a printer company, our client brought us in early in the development stage to develop a reliability program. The first step in this program was to work with their marketing and engineering teams to develop complete reliability goals for the product. Once the teams all agreed on the goals, we began working with each team to apportion the system goals to each assembly. After we completed the Reliability Apportionment, each team started the design process with a clear understanding of the goals to which they had agreed.

We worked with our client throughout the reliability program. After our client started shipping the product, we set up a measurement system to compare the field data to our original goals, and the goals proved to be very accurate.

7 Benchmark against Your Competition

Benchmarking is the process of determining and comparing reliability-related metrics for a set of specific products in a specific market. The purpose of Benchmarking is to gain a clear understanding of how your product measures up to your competitors' product in your market. Benchmarking is crucial both to a start-up company as well as an established company that is coming out with a new product to ensure that the new product is competitive based on reliability and cost.

Benchmarking is often useful even if your customers have specified the reliability requirements, as this provides you with a "sanity check" against the rest of the industry. To benchmark your product effectively, you will need to work with your marketing department. They know who the competitors are, and they know what your customers are asking for in terms of reliability.

7.1 Three Forms of Benchmarking

There are three main forms of benchmarking:

1. Product Benchmarking
2. Process Benchmarking
3. Competitive Analysis

7.1.1 Product Benchmarking

Product Benchmarking entails comparing product requirements such as:

1. Mean time between failure (MTBF)
2. Dead on arrival (DOA) rate
3. Annualized failure rate (AFR)
4. Availability
5. Maintainability

7.1.2 Process Benchmarking

Many of our clients tell us that it is nearly impossible to benchmark products against their competitors because their competitors' yield figures are tightly guarded secrets. If this is the case, then I recommend you resort to Process Benchmarking and find out what processes they use in developing their product. Process Benchmarking entails comparing process methodologies, such as in-house versus outsource builds, quality philosophy, and screening methods. Perhaps this won't be direct input for your goal statement, but it will certainly play a part in what you put in your Reliability Program Plan (RPP).

7.1.3 Competitive Analysis

The third type of Benchmarking technique is called Competitive Analysis. With this technique, compare your product's reliability performance to competitive products. Use the results as input to your Gap Analysis to determine appropriate next steps to improve your organization's reliability program.

The two main types of Competitive Analyses we recommend are:

1. Competitive HALT
2. Competitive Teardown Analysis

The order in which you execute these two techniques matters only if you have a small sample size. If you have a small sample size and can easily reassemble the product after the teardown analysis, then it makes sense to perform the Competitive Teardown Analysis first. Otherwise, it makes sense to perform the Competitive HALT first.

7.1.3.1 Competitive HALT

With a Competitive HALT, subject two or more products to HALT up to their operational/destruct limits and then compare the product margins between the products. With like technologies, a product that has better margins is generally a more reliable product.

7.1.3.2 Competitive Teardown Analysis

With a Competitive Teardown Analysis, compare two or more products by disassembling each and then comparing a number of different attributes. There are two key elements to this process:

1. Develop a meaningful set of attributes.
2. Develop an objective scoring system for each attribute.

7.1.3.2.1 Set of Attributes

Here is an example of a set of attributes you can use during Competitive Teardown Analysis for electronic and mechanical products:

1. Manufacturability
2. Inherent reliability
3. Human factors/safety
4. Maintainability/serviceability
5. Mechanical design
6. PCB design
7. Other design considerations

7.1.3.2.2 Scoring System

Depending on the number of different products involved, develop a customized scoring system to differentiate the best product from the worst product. We typically use a one to five scoring system with five being the best. If both products are very good and one product is only slightly better than another product, you can score the first product a five and the second product a four. However, if one product is significantly better, you can score the first product a five and the second product a one. Here is an example of scoring system you may use.

Table 7.1: *Example of a Competitive Teardown Analysis Scoring System*

Score	Description
1	Worst, missing most elements of attribute being measured
2	Marginal, has some of the elements of attribute being measured
3	Good, has many of the elements of attribute being measured
4	Very Good, has most of the elements of attribute being measured
5	Exceptional, has all of the elements of attribute being measured

RELIABILITY INTEGRATION: Integrating Competitive Analysis with HALT
We often use HALT as a Competitive Analysis technique because HALT can quickly differentiate the reliability of two different products. There is more to the reliability of a product than can be discovered in HALT alone (see the list in Section 7.1.3.2.1), but HALT is a good quick litmus test.

CASE STUDY: Performance Benchmarking
A computer peripherals company was competing in the area of reliability, and their marketing department asked us to help with setting up Performance Benchmarking goals, focusing on elements of technical quality and reliability as well as features. Our client gave us a list of their competitors. We reviewed our client's product and its usage and came up with Performance Benchmark metrics. Through internet searches, trade magazines, and research reports, we found the necessary reliability data for our client's competitors. With this information, we helped our client develop clear and competitive goals that they used throughout their development process.

8 A Gap Analysis Will Tell You Where to Focus Early

A **Gap Analysis** naturally flows from the Benchmarking Analysis. Once you complete the Benchmarking Analysis, you should then compare the results with your current capabilities, and this becomes the input to the Gap Analysis. The larger the gap, the more effort you will need to put forth to meet your requirements.

The objective of a Gap Analysis is to measure the gap between where you currently are with your reliability goals compared to where you have set your reliability goals to be.

The Gap Analysis will also show how far away you are from achieving your goals. The Gap Analysis becomes the basis for identifying implementation actions and priorities. Knowing the size of the gap for each particular reliability and quality metric is valuable because it will then tell you how much resources you will need to dedicate in order to meet these metrics.

In your Reliability Program Plan (RPP), you should describe your reliability goals for the project and what reliability techniques you will use to achieve these goals. The Benchmarking results and the Gap Analysis results are two of the best sources of information in establishing this. A good Benchmarking exercise can reveal not only the reliability results of other products in a similar industry, but also the techniques used to achieve these results.

Measure the gap for each of these metrics by comparing the goals to your current performance level. You can then input this directly into your RPP so you can specify the reliability techniques and tasks that will be most effective at reducing or eliminating your gap.

RELIABILITY INTEGRATION: Integrating Gap Analysis with the Reliability Program—The Size of the Gap Determines Which Technique to Use

If your gap is small, then you may only need to make minor adjustments to your reliability program, or you may need to provide some training in a particular area. However, if the gap is much larger, then it is likely you will have to perform some analysis or experiments to determine the cause of the failures, which may require that you utilize techniques that you hadn't previously deployed, such as Finite Element Analysis (FEA) or Design of Experiments (DOE).

CASE STUDY: Gap Analysis Drove a Major Redesign

We performed a Reliability Assessment followed by a Gap Analysis for a semiconductor manufacturing company. We found that our client needed to improve their reliability significantly to be competitive in the market. We worked with them to embark on a reliability improvement process consisting of increasing the effectiveness of some existing techniques and introducing a few new techniques. One year later, they had us return to reassess their process, and we found that they were meeting their goals.

9 Metrics (The "Are We There Yet?" Technique)

What **Reliability Metrics** will you use to measure that you are meeting your goals? Reliability Metrics provide the measurements and milestones, the "are we there, yet?" feedback that your organization needs to ensure you are on track toward meeting your goals. Without these metrics, you have no way of knowing if you are on track. Once you have established your goals, define the key metrics you can use to monitor your reliability goals.

Examples of Reliability Metrics during development:

- Predicted Prediction results
- Reliability Demonstration Test (RDT) results
- Accelerated Life Test (ALT) results
- Vendor test data
- Software defect density

Examples of Reliability Metrics after deployment:

- Field failure rate
- Warranty return rate
- Annualized failure rate (AFR)
- Dead on arrival (DOA) rate
- No problem found (NPF) rate
- Software bug rate

9.1 Reliability Metrics Provide a Means to Monitor Reliability Performance

A set of reliability goals and metrics provides boundaries for specific product reliability objectives, permits comparisons of predictions and reliability testing results to business needs, and enables the cascading of reliability objectives to key vendors. Most importantly, reliability goals and metrics provide your organization with a means to monitor reliability performance.

9.2 Reliability Metrics and the Reliability "Bathtub" Curve

When setting goals and metrics, you should consider each of the three different phases of the Reliability "Bathtub" Curve or product life cycle—infant mortality, steady state, and wear-out.

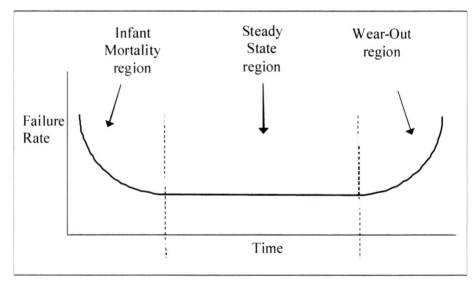

Figure 9.1: Reliability "Bathtub" Curve—The Ideal Version

9.2.1 Infant Mortality Region

The infant mortality region, as shown in Figure 9.1, is the region of decreasing slope, or decreasing failure rate, from time t=0 and forward. Typically the probability of failure is highest immediately after you ship the product to your customer. After that point, the failure rate reduces. However, you should be careful with this terminology because it is possible that certain circumstances don't follow this pattern. For example, a solder joint or a component may fail many months or even a few years after installation, even though the failure was related to a factory defect.

Each component in your product has an infant mortality distribution curve, and this curve can stretch well into the steady state region. By the time the steady state region begins, the infant mortality rate for these failures is well below the steady state failure rate. In Figure 9.1, we show it as going away completely even though it is still present. See Figure 9.2 for a more realistic view of the infant mortality portion of the Reliability "Bathtub" Curve. Each curve represents the dominant failure mechanism for a particular component. That component is used at a specific location in every product. If you take the entire population of products and run a test until all of those components failed and then plot the failures in time, it would take the shape of the curve. It is nearly impossible to perform this exercise, so instead we display a theoretical view of this in Figure 9.2.

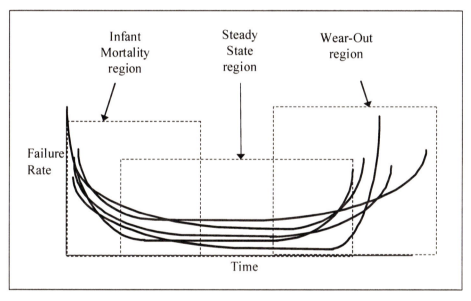

Figure 9.2: Reliability "Bathtub" Curve—The Realistic Version

9.2.1.1 Infant Mortality Examples

1. Fabrication process damage
2. Oxide defects and damage
3. Ionic contamination
4. Package defects (cracking)

5. Solder defects
6. Screws/cables not installed properly

9.2.2 Steady State Region

The steady state region, as shown in Figure 9.1, implies a constant failure rate. In this region, the probability of something failing stays relatively constant. Failures could be caused by random events such as power surges, lightning, earthquakes, or any other random occurrence. This phase doesn't have a unique starting and stopping point. Steady state failures can occur any time during the product life cycle, but these failures are most noticeable during the steady state region because infant mortality failures have reduced below the steady state rate, and wear-out failures haven't yet started to increase to the steady state failure rate.

Each component in your product has a steady state distribution curve, and this curve can stretch into both the infant mortality region as well as into the wear-out region. The steady state failure rate within the infant mortality region is typically much lower than the infant mortality failure rate and the wear-out failure rate. In Figure 9.1, we show the steady state failure rate as going away completely in these two regions even though it is still present. See Figure 9.2 for a more realistic view of the steady state portion of the Reliability "Bathtub" Curve.

9.2.2.1 Steady State Failure Examples

1. External random occurrences (lightning events, electrostatic discharge (ESD) events, power spikes)
2. Cosmic rays
3. Random event caused by user misuse

9.2.3 Wear-Out Region

The wear-out region, as shown in Figure 9.1, is the region of increasing slope or increasing failure rate. In this region, components start wearing out. Every single component in your product can wear out. However, only a few have what we call dominant wear-out mechanisms, which are failure mechanisms that will show up during the useful life of your product. A well-made solder joint, for example, will eventually wear out, but in most benign usage applications it isn't likely to do so for many years until after your product becomes obsolete. The key is to determine which wear-out mechanisms will affect your product during its useful life, and either design them out of your product or develop Preventive Maintenance schemes.

Each component in your product has a wear-out distribution curve, and this curve can start as far back as the infant mortality or steady state region. The wear-out failure rate within the infant mortality region is typically much lower than the infant mortality failure rate and the steady state failure rate. In Figure 9.1, we show the wear-out failure rate as not starting until after the steady state failure rate, even though it is present prior to this. See Figure 9.2 for a more realistic view of the wear-out portion of the Reliability "Bathtub" Curve.

9.2.3.1 Wear-Out Examples

1. Metallization Failures
    a. Dendrite growth (silver, tin)
    b. Electromigration
    c. Corrosion (copper, aluminum)
    d. Fatigue & fretting (solder)
2. Lubricant breakdown
3. Time dependent dielectric breakdown—thin oxides
4. Electrolyte loss for electrolytic capacitors
5. Mechanical wear

RELIABILITY INTEGRATION: Integrating Metrics into Your Reliability Program—Continuously Review Your Metrics
No matter where you are in your reliability program, don't lose track of your metrics. If you hear people start to talk about release dates, check your metrics and make sure you are on track to meet your reliability goals. If you are currently performing a Reliability Demonstration Test (RDT), extrapolate to determine when you will reach your goals.

CASE STUDY: MTBF as a Metric

For a telecommunications company, our client was using MTBF as their key metric during development and when they fielded the product. Several times they were surprised with spikes in the failure rate in the field and were not quick enough to react. Their metric of MTBF was not able to identify the issue. We worked with them to change from MTBF to Reliability. The result was that they were able to identify field issues much quicker because they were analyzing the failures as a function of time. When they had a batch failure with one particular component, their infant mortality rate spiked up to 5%. They identified this within two days of when the failure started occurring and were able to put in place a corrective action by the end of that same week. Using MTBF as a metric, they likely would have missed this failure entirely. See the Reliability Integration section of Chapter 49 for more details on the differences between MTBF and Reliability as a metric.

10 Break the System Up into Blocks

Once you have stated the overall reliability goal, now it is time to apportion (or allocate) the goal down to the different assemblies within a product. We call this **Reliability Apportionment** or **Reliability Allocation**. It is only by stating goals at the assembly level that you can really get started with a design.

Let's revisit the goal statement example from Chapter 6:

A desktop business computer in an office environment with 95% reliability for one year.

For simplicity, consider five major elements of the computer:

1. Motherboard
2. Hard Disk Drive
3. Power Supply
4. Monitor
5. Keyboard

10.1 Reliability Apportionment Process

Reliability Apportionment is a three-step process:

1. Come Up with Initial Estimates
2. Determine the Reliability of Each Assembly
3. Re-Apportion as Needed to Optimize System Goals

10.1.1 Come Up with Initial Estimates

First, draw out a Reliability Block Diagram and give each assembly the same goal as the others. If failures of each assembly cause a system failure, then they are independent (this is called a "series" model). The simple multiplication of the reliabilities of each assembly should result in meeting the system goal. Figure 10.1 illustrates the initial step for Reliability Apportionment where we allocate 0.99 for the reliability of each assembly.

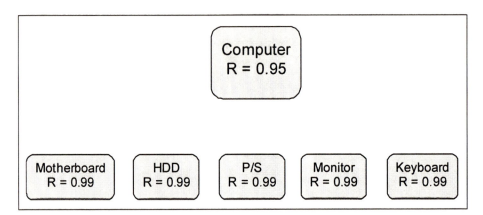

Figure 10.1: Reliability Apportionment Step 1: Initial Allocation

$R_{Motherboard} \times R_{HDD} \times R_{P/S} \times R_{Monitor} \times R_{Keyboard} = R_{System}$ *Formula 10.1*

or

0.99 x 0.99 x 0.99 x 0.99 x 0.99 = 0.95

Note that this is just a starting point assuming we have no historical information or vendor data.

10.1.2 Determine the Reliability of Each Assembly

There are three methods you can use to determine assembly reliability:

1. Historical data from similar products
2. Reliability estimates/test data by vendors
3. In-house reliability testing

Using these methods, you can then come up with first pass assembly reliability estimates. An example of what you may come up with is shown in Figure 10.2.

Chapter 10: Break the System Up into Blocks

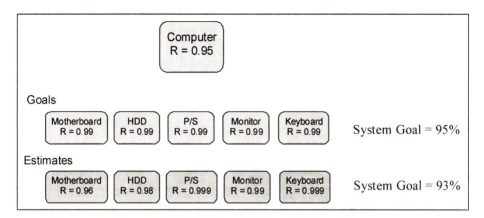

Figure 10.2: Reliability Apportionment Step 2: First Pass Assembly Estimation

10.1.3 Re-Apportion as Needed to Optimize System Goals

Notice that the first pass estimates don't meet the system goal. Now what? You must resolve the gap! Take each individual assembly and determine what is causing the gap between the current estimated reliability results and where you desire the reliability results to be.

The motherboard initial allocation goal was 99%, and you have estimated the goal at 96%. This is the largest gap and the lowest estimate. First, will addressing the known issues bridge the difference? If not, then use FMEA to populate the list of what to fix. Next, you should validate your reliability improvements. This information all goes into your Reliability Program Plan (RPP).

The hard disk drive (HDD) initial allocation goal was 99%, and you have estimated the reliability to be 98%. This is a small gap and there is a clear path to resolution. HDD reliability and operating temperature are usually related. By lowering the internal temperature, the HDD reliability will improve. When the relationship of the failure mode and either design or environmental conditions exist, you don't need FMEA. You can go straight to design improvements. You can then use Accelerated Life Testing (ALT) to validate the model and/or design improvements.

The power supply initial allocation was 99%, and you have estimated the reliability to be 99.9%. The estimate exceeds the goal. Further improvements aren't cost effective given their minimal impact to system reliability. In fact, it may be possible to reduce the reliability (select a less expensive power supply) and use your cost savings to improve the motherboard reliability. For any assembly that exceeds the reliability goal, explore potential cost savings by reducing the reliability performance. This is only done when you have accurate reliability estimates.

You should continue to do this for each assembly within your system. Note that you can now take all of this information and enter it into your RPP.

10.2 Progression of Estimates

When estimating the reliability, there are a number of different sources for the data, and some are more accurate than others. Figure 10.3 shows the progression of estimates from least accurate to most accurate. Unfortunately, the more accurate the data, the harder it is to find the data early in the product development cycle. Therefore, you are usually left with starting with the less accurate data and then improving the data accuracy as the project progresses.

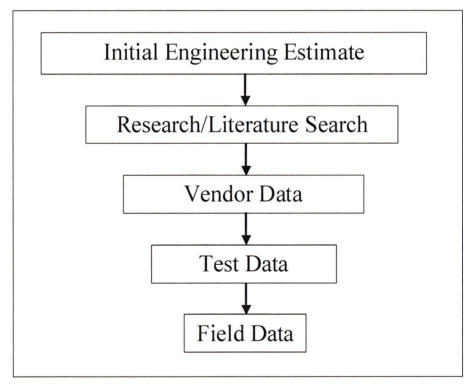

Figure 10.3: Progression of Estimates

RELIABILITY INTEGRATION: Integrating Apportionment with a Reliability Plan—Apportionment Drives the Reliability Plan
How can you write a good RPP unless you have an idea of the level of reliability of each of the components that are going into the product? If you lack this data at the beginning of the program, then you should apportion the reliability evenly across each assembly. Next, you should start gathering this data from research, vendor data, and field data on previous products and combine it into a Reliability Apportionment diagram as shown in Figures 10.1 and 10.2. Let the gaps drive which reliability activity is most suitable.

CASE STUDY: "Let's Just Set a Goal of 98% for All our Assemblies"
We were working with an airplane equipment company, and they decided to set the goals for every module at 98% reliability. The problem was that each module had varying degrees of complexity (some differed by an order of magnitude based on component count). When they started collecting field data, they were frustrated that half of their assemblies were exceeding their goals, and half weren't. We came in and helped them apportion their reliability more appropriately, taking into account both complexity and past history of components. The result was that we developed much more realistic goals for each assembly and were able to predict field performance more accurately.

11 Golden Nuggets Are the Golden Rule

Golden Nuggets refers to those few techniques that your organization does well, so well in fact that these techniques become engrained into your culture. The Golden Nuggets become part of your "secret sauce" that give your product and company a competitive advantage in specific areas. Sometimes your organization doesn't know that they do them well or even know that they are doing them at all. It is your responsibility to point out these Golden Nuggets to your organization because you should always reinforce good behavior.

When assessing your organization, look for these Golden Nuggets. Perhaps your organization has a keen ability to use mechanical simulation techniques such as Finite Element Analysis (FEA) or statistical data analysis techniques such as Design of Experiments (DOE). If you discover this, make sure to point it out to your organization and help them use these Golden Nuggets to their advantage.

Other times, you may discover that your organization has an excellent ability to work as a team and to reach a consensus. Perhaps they don't even recognize that they possess this skill. Again, you should encourage this behavior when you recognize it.

Having a well-defined (and well-followed) product development process could be another example of a Golden Nugget. Many times we have seen companies with disjointed product development processes, and when we followed failures back to a root cause, we often found that they bypassed a process. Consistently following a product development process is a key part to reducing failures, and the companies we have worked with that already had this in place were typically the companies with more reliable products.

RELIABILITY INTEGRATION: Integrating Golden Nuggets With Your Lessons Learned Process
At the end of the development program, review the results of your program to find areas of opportunity as well as areas of your program that went well. Compare your analytical techniques to your test results to determine how accurate your analyses were at predicting reliability test failures. Compare your test results to your field results to determine how accurate your reliability tests were at predicting field failures. Facilitate a meeting with the team to uncover inefficiencies in each process and to discover processes that worked well. Capture areas of opportunity in your Lessons Learned database. Capture areas of your program that worked well and make them part of your Golden Nuggets.

We interviewed a few best-in-class companies and asked what they thought separated them from their competition. Most felt that it was their Golden Nuggets. I won't mention any particular techniques here because it isn't the technique that matters. What matters is how the technique is used. Any technique can be used well and it can also be used poorly. However, if the technique is an integral part of your reliability program and it is performed well, it can help separate you from your competition.

CASE STUDY: Golden Nuggets Drive Best in Class Companies
We performed a Reliability Program Assessment for an aircraft manufacturer. During the assessment, we discovered that they had a great method of collecting data and feeding it back to their design process. During our recommendations process, we highlighted this and made sure that as we helped them implement changes to their processes, we didn't make changes to this Golden Nugget because it was working well. The end result was that they ended up with a much stronger reliability program with the capability of delivering reliable products.

12 You Need a Plan

A **Reliability Program Plan (RPP)** ties together customer requirements, business opportunities, and employee opportunities. For customer requirements, your RPP ensures that you meet the terms of your contract with your customers and that you meet customers' expectations. For business opportunities, a good RPP can help reduce expenses and improve brand perception. For employee opportunities, a good RPP can provide employee direction and empowerment.

An RPP is crucial at the beginning of the product life cycle for each product. During your reliability program, you should update this plan as you discover more information about the product. In this plan, you define the following eleven key steps:

1. What Are Your Reliability Goals?
2. How Will You Apportion Your Goals to Individual Assemblies?
3. What Are Your Historical Issues for Reliability?
4. What Are Your Constraints in the Program?
5. What Are Your Critical to Quality (CTQ) Parameters?
6. Which Reliability Techniques Will You Use?
7. How Will You Implement and Integrate Each Technique?
8. What Metrics Will You Use to Measure If You Are Meeting Your Goals?
9. What Is the Frequency in Reporting Results and to Whom?
10. Assess If You Need to Change Any Goals
11. What Is Your Schedule for Meeting These Goals?

12.1 What Are Your Reliability Goals?

Your plan should start with a goal statement. See Chapter 6 on Goal Setting for more details.

12.2 How Will You Apportion Your Goals to Individual Assemblies?

Next, you should allocate your goals down to the individual assemblies. To do this, you should take into consideration what has been the past performance of the product and what is the gap between past performance and current assembly goals. See Chapter 10 on Reliability Apportionment for more details.

12.3 What Are Your Historical Issues for Reliability?

To assist in selecting goals for the quality and reliability metrics, perform a historical review of return data for the similar predecessor products. If your organization is building its first product, then you may skip over this section. If not, then review your field data as well as your test data from previous products to determine if there were any major issues. If you are keeping a Lessons Learned database, pull it out and review.

12.4 What Are Your Constraints in the Program?

Next, you should state any constraints that you may have that would prevent you from meeting your goals. List how you plan to overcome these constraints. There are usually ways around each type of constraint, but you need to know this during the planning stage. State the constraints as well as how you are going to compensate for the constraints in your approach. Otherwise, people reading the plan will question the approach.

Here is a list of a few of the constraints you may have:

- Time constraints: Perhaps there isn't enough time in the program to complete all of the reliability tests you have planned.
- Money or budget constraints: Perhaps there isn't enough money in the budget for the number of samples you need for testing.
- Resource constraints: Perhaps you are understaffed for this particular program.
- Pre-determined methods or techniques: Perhaps your customer dictated specific types of analyses or tests you must perform.

- Pre-determined vendors: Perhaps your purchasing department has already chosen the contract manufacturer (CM) or another key supplier.

12.5 What Are Your Critical to Quality (CTQ) Parameters?

CTQ parameters are specific, measurable characteristics of a product or process that are necessary for your customers' satisfaction. Establish your CTQ parameters during the RPP and then add to these during the Failure Modes and Effects Analysis (FMEA).

When adding CTQ's from an FMEA, use the failure modes with the highest RPNs as well as the results from previous reliability test results to identify your CTQ parameters that you should flag for further evaluation and tracking.

Some CTQ parameters will apply to variables you can control or monitor at your manufacturing site, while others your component supplier will control or monitor. You should determine your CTQs to ensure you meet the reliability targets. This is the shared responsibility between your design team, product assurance, quality assurance, and manufacturing. You should also monitor your CTQ parameters to ensure that your product stays within a well-defined process window and you minimize customer complaints. Prior to the pilot run, preferably before pre-production, hold a design review to focus on each CTQ in turn to define a suitable monitoring plan. Where possible, apply statistical process controls to monitor your CTQs.

12.6 Which Reliability Techniques Will You Use?

Based on the size of your gap (from your Gap Analysis), and what is causing this gap, you need to choose which reliability techniques to implement. If the gap is large, you will need to invest a lot of resources in the design techniques prior to building and testing prototypes, such as:

- Design of Experiments (DOE)
- Failure Modes and Effects Analysis (FMEA)
- Derating Analysis

If your gap is small, you should invest more resources in the prototype techniques, such as:

- Highly Accelerated Life Test (HALT)
- Reliability Demonstration Tests (RDT)
- Accelerated Life Tests (ALT)

If your gap is largely a result of production escapes, you may want to invest more effort into developing good manufacturing reliability techniques, such as:

- Highly Accelerated Stress Screens (HASS)
- Ongoing Reliability Tests (ORT)
- Out-of-Box Audits (OOBA)

As with most programs, the gap will likely be large in some areas and small in others. Therefore, you should develop a well-balanced reliability program and select techniques from each of the phases: design, prototype, and manufacturing. Your RPP is very much dependent on the situation; as a result your plan may change from product to product depending on many factors.

12.7 How Will You Implement and Integrate Each Technique?

The implementation and integration of each technique is perhaps the most difficult to plan. Here you should estimate the effects each technique will have on the overall reliability to understand how you will close the gap.

For this, you should look at specific issues that occurred on previous products. You need to understand how a specific technique will help mitigate each issue on the next generation of product. If you can also quantify the reduction that you can achieve using the most suitable technique, then you have evaluated how you are going to close the gap.

Example: *A power supply company's current product is running at a 0.25% dead on arrival (DOA) rate (this is the rate of products that don't work when your customer first receives and installs the product) per month. Their goal is to reduce this by 50%.*

In this example, the DOAs tend to focus around solder issues. For this next generation, they decide to choose HASS as their technique to solve this. Through research, they learn that HASS is about 90% effective in finding and preventing solder defects from escaping into the field. Therefore, they write in their RPP that they expect to meet and exceed their 50% reduction target.

However, you aren't finished. You need to describe how you will implement and integrate each technique. To say that you will use HASS and you have evidence it will work is only the first step. In the next step, you should explain how you will implement it successfully. In your RPP, you need to:

- Determine what level HASS will be performed (assembly or system).
- Outline functional and environmental equipment needed.
- Determine production needs and throughput.
- Understand manpower needs.

Are you done? Not quite. Next, you should explain the integration. What techniques will feed into HASS in order to make it successful? How will they be used? Here are examples of techniques that feed *into* the HASS process:

- Using a Reliability Prediction, you can determine how much margin you have against your reliability specification, but only for design reliability, not for manufacturing reliability. The less margin you have in your design, the more effective of a HASS you will need to implement.
- Using FMEA, you can understand technology limiting components. This is important when developing HASS to ensure that you don't choose stresses that wear out the product prematurely.
- Using HALT, you can develop margins. This is important when implementing HASS because the more margin you have in your design, the more robust of a manufacturing screen you can develop during HASS.

Next, you should determine what techniques will HASS feed into? Here are examples of techniques that feed *from* the HASS process:

- You can use your Field Data Tracking System to monitor your DOA rate.
- You can use your repair depot to reduce your no problem found (NPF) rate. NPFs are instances in which your customer complains of a problem, but you can't duplicate the problem. In these situations, you can develop a repair depot environmental screen that stresses the product enough to find the failure if indeed there was a flaw. This can be the same screen that you use during HASS.

12.8 What Metrics Will You Use to Measure If You Are Meeting Your Goals?

Next, you should choose which Reliability Metrics you will use to measure where you are in comparison to your Reliability Goals.

12.9 What Is the Frequency in Reporting Results and to Whom?

You should decide what information to report (typically in the form of the metrics you have chosen), how often to report results, and to whom you will report the results. As part of the reporting process, if there are areas of the program that aren't working or are behind schedule, you should develop a plan to get the program back on track.

12.10 Assess If You Need to Change Any Goals

Occasionally you will find that the goals you started with are not correct and you need to change them. The reason for this may be because something changed in the industry (perhaps one of your customers asked for a change of goals or perhaps the competition changed), or you made an error in assessing the reliability of a particular technology making it impossible to meet your goals within your budget, or with the current technology available. Whatever the reason may be, you should update your reliability goals and update your RPP.

12.11 What Is Your Schedule for Meeting These Goals?

The last piece of your RPP is the schedule. With an infinite amount of time (and money) you can achieve almost any level of reliability. However, you won't have this luxury! You should schedule your reliability activities and ensure that they are aligned with the schedule for the overall program.

First, determine the order of occurrence of the techniques. If you were thorough in describing the techniques and the integration of each, then this should be straight-forward. Next, estimate a length of time that you will use each technique. Then, plot these on a timeline along with dependencies.

Finally, you should compare this timeline with the master project schedule and make adjustments as necessary. If you can't complete a specific reliability task prior to the product's release, you need to develop a workaround. In some cases, you won't have enough time to reach your reliability goals during an Accelerated Life Test (ALT) or a Reliability Demonstration Test (RDT). If this is the case, you may want to lower the confidence level required (the confidence level tells how likely the test results will match the field results) at the time the product is released into the market, then continue the testing and gather data to gain the necessary confidence in your product's demonstrated reliability number. You should do this before you ramp up the volume of this product. Note that this will only work if the volume ramp isn't immediate. For products that have an immediate ramp to volume (such as products that are released just prior to Christmas), you should achieve your reliability goals before its market release.

RELIABILITY INTEGRATION: Integrating the Reliability Plan into a Reliability Program—The Reliability Program Integration Plan
The industry refers to this type of plan as the Reliability Program Plan (RPP). I prefer to call it the Reliability Program Integration Plan (RPIP) because if it is written and executed well, it will integrate all of the different activities in your program together, starting with the goal statement all the way to collecting field results. Such a plan will determine whether or not you met your goals.

CASE STUDY: The Reliability Case

We were working with a computer company out of the United Kingdom. They asked us to help them write their Reliability Case. (not a Reliability Plan but rather a Reliability *Case*). The concept of a Reliability Case was developed in the UK and is becoming much more common around the world. The Reliability Case requires you to "prove" your Reliability Plan. You can't just write a plan. You must prove it with evidence in a Reliability Case.

The principal goal of the Reliability Case is for the supplier to guarantee that their product will meet an agreed set of in-service reliability requirements. The onus of responsibility is on the supplier to build the case by gathering evidence showing that the product will meet the reliability requirements. The supplier then develops a Reliability Case Report, which contains a summary of the Reliability Case with supporting evidence.

This practice is starting to be introduced in the USA as well. We've seen this used on a number of military contracts, but many industries can benefit from this approach. We are now recommending this process to many of our clients and have seen dramatic improvements in our clients' ability to meet their reliability goals.

Much of this seems to have been borrowed from the legal profession. We often joke about how litigious society has gotten. However, following the legal path of moving to a Reliability Case makes a lot of sense, and I see the trend moving more and more in this direction.

Part III
Design Phase

CONCEPT

DESIGN

PROTOTYPE

MANUFACTURING

RELIABILITY
INTEGRATION[SM]

Reliability Engineering Services Integrated Throughout the Product Life Cycle

13 The Precious Diamond of Product Development

Engineering teams should balance project costs, customer maintenance costs, quality, schedule, performance, and reliability (and possibly other factors specific to your industry) to achieve optimal customer satisfaction. In Figure 13.1, I will use four key factors—reliability, schedule, performance, and cost, and reliability—to illustrate the product development decision process.

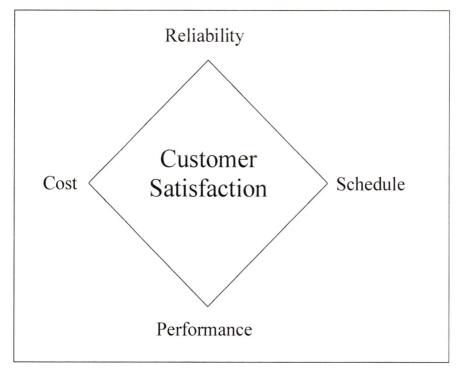

Figure 13.1: The Product Development Decision Diamond

When we meet with an engineering team to talk about reliability, the first question we ask is "Is reliability important to your company?" What do you think their response will be? Of course they say that reliability is important. The next question we ask is, "If you had to rank reliability with the other three characteristics of cost, performance, and schedule, where would you rate reliability in the order of priority?" Guess what is the most common answer? You guessed it—at the bottom. Why is that? It is usually because they believe that they have to complete the design and get the product working before they can worry about reliability. The problem with this approach is that once they complete the design, it is difficult and expensive to improve the reliability.

This is the classic "throw it over the wall" approach to reliability. Wait until you complete the design, then give the product to the reliability group so that they can find all of the problems through product testing. This method won't get you a reliable product.

The better approach is to take all four of these parameters and use them in every decision you make. This doesn't mean that reliability will be the key parameter you use during every decision, but it does mean that you should include reliability in each decision. For example, you should choose the more reliable component if the reliability of one component is significantly better than the alternate component and if there is little difference in performance, cost, and schedule. On the other hand, you should choose cost if there is a big cost difference and if there is little difference in the other attributes.

RELIABILITY INTEGRATION: Integrating the Product Development Decision Diamond into Day-to-Day Decisions
The best way to use the product development decision diamond is in day-to-day decision making in a reliability program. Your purchasing group is usually going to push you to use the least expensive materials or to use vendors with the best payment or delivery terms. If this happens, you need to push back and analyze each situation against the different decision parameters—project costs, customer maintenance costs, quality, schedule, performance, and reliability. You may face these decisions anywhere in the product development process, not just at the beginning.

CASE STUDY: The Five Cent Savings

We were working with a power supply company, and during the testing process, one of the buyers found a component that cost five cents less than the previous component. They wanted to switch to this new component. We first compared the specifications of this new component to the original component. All of the parameters appeared to be comparable. We then installed each component into two separate assemblies and performed a side by side HALT. Through the HALT, we determined that the new component was far inferior to the original component; therefore, we recommended against using this new component. This five-cent savings could have cost our client millions of dollars. Our client agreed with our recommendation and didn't use the new component.

14 It Is All a Balancing Act

When you start writing your Reliability Program Plan (RPP) and start building your reliability program, there are different philosophies you can use as well as different reliability techniques. In this chapter, we will explore some of your different options.

14.1 Three Philosophies to a Reliability Program

When building a reliability program, you can use three philosophies:

1. Build, Test, Fix Approach
2. Analytical Approach
3. Balanced Approach

14.1.1 Build, Test, Fix Approach

With the Build, Test, Fix Approach, you will spend less time and effort using reliability analysis techniques and will quickly get the product into reliability testing. An example of a company that uses this approach is a manufacturer of MP3 players or some other relatively inexpensive commodity item. For these types of products, it may make more sense to rely heavily on testing.

14.1.2 Analytical Approach

With the Analytical Approach, you will spend more time and effort using reliability analysis techniques and less using reliability testing techniques. An example of a company that uses this approach is NASA. When they develop products such as the space shuttle, they typically make very few test samples because of the high cost of each sample. In this case, more analysis would be prudent.

14.1.3 Balanced Approach

With the Balanced Approach, you balance your time and effort between analysis and testing. An example of a company that uses this approach is a medical product company. They will want to balance the analysis with the testing because both are important.

The reliability goal and the gap between your goal and your current capability (Gap Analysis) will guide you whether you should use more of a Build, Test, Fix Approach, more of an Analytical Approach, or a balance between the two.

14.2 Different Approaches to Reliability

There are two fundamental approaches to reliability:

1. Reliability Measurement Approach
2. Reliability Improvement Approach

A successful reliability program requires a good combination of the two approaches.

14.2.1 Reliability Measurement Approach

Reliability Measurement Approach uses reliability techniques designed to *measure* where you are compared with your reliability goals. Examples of these techniques include the following:

- Reliability Predictions
- Accelerated Life Tests (ALT)
- Reliability Demonstration Tests (RDT)
- Ongoing Reliability Tests (ORT)

It is possible to make improvements to your product based on the results of these techniques, but that isn't the focus. For example, if you perform an RDT, it is possible that you will experience a failure; through failure analysis and corrective action, you may make an improvement to the product. However, in a typical RDT, you may only experience one or two failures; therefore, there is little opportunity for improvement.

14.2.2 Reliability Improvement Approach

Reliability Improvement Approach uses reliability techniques design to *improve* the reliability of your product. Examples of these techniques:

- Failure Modes and Effects Analysis (FMEA)
- Highly Accelerated Life Tests (HALT)

It is possible to measure where you are based on the results of these techniques, but that isn't the focus. For example, if you perform an FMEA, you can use a risk priority number (RPN) scoring system to measure where you are, but it is difficult to tie this back to a reliability metric.

RELIABILITY INTEGRATION: Integrating Reliability Measurement and Improvement Techniques—Integrating HALT and the HALT Calculator
For years, engineers have asked the question "Is it possible to calculate a mean time between failure (MTBF) number from the results of High Accelerated Life Testing (HALT)?" My classic response was "No, but you can use the results of HALT to develop an RDT." Once you discover the product margins in HALT, you can then develop a more intelligent RDT by increasing the RDT stress levels and gaining more acceleration. However, many engineers weren't satisfied with this approach because it required another test. That is why we developed a HALT-to-AFR Calculator. Using this calculator, we can calculate the actual failure rate of a product directly from the results of HALT. This is a good example of integrating the Reliability Improvement Approach technique of HALT with the Reliability Measurement Approach technique using the HALT-to-AFR Calculator. See Chapter 38 for more details on the HALT-to-AFR Calculator.

CASE STUDY: How Many Samples Do You Want?
Some industries just can't supply many samples for testing. Recently, I was at a meeting with engineers from one of our space customers in this situation. The group gave me one sample for testing and added stipulation by saying, "Test it, but not too severely because the test sample will also be used as part of our flight hardware". That definitely changed our strategy in how we approached our testing. We recommended to our client to perform more reliability analysis and to use reliability testing for areas of the analysis that had uncertainty. This approach ended up working for them. They were able to obtain the reliability information needed, and we did not damage their sample in the testing process.

15 New Technologies Bring New Risks

A **Technology Risk Assessment** is the identification, categorization, and prioritization of hardware and software risks to achieve key reliability business objectives. The objective of a Technology Risk Assessment is to identify and establish an action plan to remove reliability uncertainty and mitigate business risks involved with product technology reliability performance.

There are two areas of technology risk that can lead to an unknown amount of increased reliability risk:

1. New product concepts or new technology innovations may have an unclear reliability performance.
2. An existing product may be heading into a new market or usage profile with unknown changes in product failure rates.

You can use the method of Technology Risk Assessment to:

- Frame early product design architecture and engineering approaches to solve key reliability issues to minimize overall program risk.
- Establish realistic reliability goals.
- Provide input to supplier selection criteria and reliability management programs.

Through concept reviews and analysis, and via workshops with key designers/suppliers, you should establish expected performance, define failure, and identify critical low margin elements. In addition, look for the key technologies that may lead to design flaws impacting your product's reliability.

RELIABILITY INTEGRATION: Integrating New Technologies with the Best Reliability Techniques

Depending on the new technology being introduced, you will need to choose an appropriate reliability technique to prove that the new technology is as reliable as the previous technology. In many cases, an ALT is a good technique to use, but for it to be effective, you need to determine the failure mechanism of both technologies, and they are likely to be very different. Take the simple example of a mechanical relay compared with a solid state relay. If you put both of them into a temperature chamber and elevate the temperature, the temperature at which they fail and the failure mechanism when they do fail would likely be very different. Therefore, the resulting test acceleration would be very different. If you don't take this into account when designing your ALT, you could get very confusing and misleading results.

CASE STUDY: Integrating a New Technology Component into a Product—the Storage Industry

What happens when you want to switch to a new technology? Will this technology work as well as the one it is displacing? This is currently going on in the storage industry. Solid state drives have been closing in on mechanical drives for years in the area of price, performance, and reliability. At some point, they may overtake mechanical drives. They have already surpassed mechanical drives in harsh environment industries, such as MP3 players and mobile phones. However, in industries with more benign environments that require higher reliability, the answer isn't as clear. Some types of rewriteable memory only have a life of several thousand reads/writes before read/write errors start to occur. That's fine for Universal Serial Bus (USB) memory sticks, but what about high reliability/high read/write applications, such as banking, where you can get millions of reads/writes? As the size of the memory gets smaller and smaller, the number of read/write cycles before failure actually decreases due to failure mechanisms such as hot carrier injection and dielectric breakdown. Companies are working hard to overcome this with error correcting schemes and other methods, and someday, they may have a solution. However, until they do, some industries can't make the switch.

16 Think How Your Product Will Fail: The FMEA Technique

Failure Modes and Effects Analysis (FMEA) is a systematic technique of identifying and preventing product and process problems *before* they occur. With FMEA, you explore potential failure modes, assign a resulting risk level, then prioritize actions by risk level. Risk management is a process for identifying hazards associated with a product, estimating and evaluating the associated risks, controlling these risks, and monitoring the effectiveness of the control. The process includes risk analysis, risk evaluation, and risk control. Risk management uses FMEA as a technique when evaluating and controlling risks.

Note that many people use the term FMECA instead of FMEA. In the acronym FMECA, the "C" stands for criticality. The process is the same except you capture one more score—the criticality of the failure. See military standard MIL-HDBK-1629 (which can be found on the Department of Defense website http://dodssp.daps.dla.mil) for a more detailed explanation of the term.

Often, people confuse the terms "failure mode" and "failure mechanism." We use the two terms quite a bit in this book, so I will provide the clarification here. The failure mode is the actual symptom of the failure such as "failed component" or "degradation of performance." The failure mechanism is the cause of the failure mode such as "corrosion" or "vibration."

FMEAs are introduced early in the design process, then updated throughout the life cycle of a product to capture changes in the design, as well as to update the FMEA on the effectiveness of the corrective action. Early in the design, you may identify a potential failure mode and develop a corrective action for that failure mode, but you won't know how effective that corrective action will be until you actually implement the action and test its effectiveness. This usually comes much later in the program.

16.1 Different Types of FMEA's

To make an FMEA more manageable, you should first decide what type of FMEA you want to perform—Design, Process, User, and Software FMEAs are a few of the more common types.

Design FMEAs are performed on the system at the design level. The purpose is to analyze how failure modes affect the system and to minimize failure effects upon the system.

Process FMEAs are performed on the manufacturing processes. They are conducted through the quality planning phase as an aid during production. The purpose is to analyze and correct the possible failure modes in the manufacturing process, including limitations in equipment, tooling, gauges, operator training, or potential sources of error.

User FMEAs focus specifically on the end user and how they will use, misuse, or possibly even abuse the product. An input to the User FMEA is the user manual. The User FMEA will look at installation, use, and end-of-life situations. Whenever a user is involved, you should pay specific attention to the possibility of the user using the product incorrectly, risking either the integrity of the product or, worse, creating an unsafe situation.

Software FMEAs focus on potential software bugs as well as errors in interfaces and errors in boundary conditions. This is an excellent technique if you have a set of bugs and are trying to determine the likely cause.

You should also decide if you want to perform the FMEA at the piece-part level or functional level. At the piece-part level, start with each individual component (if you are performing a Design FMEA), process step (if you are performing a Process FMEA), user step (if you are performing a User FMEA), and software subroutine (if you are performing a Software FMEA). Then, identify all failure modes with each component, process step, or software subroutine.

At the functional level, identify the major functions of the product and identify failure modes for each of these functions. For complex systems, we often start at the functional level and then only go down to the piece-part level if at the functional level we identify a failure mode with a high risk. This can save a lot of time and expense.

16.2 Different FMEA Standards

There are at least 20 different standards and guidelines for FMEAs. Most are very similar in methodology and differ only in how to assign a value to a failure mode. In addition, most of these only outline the actual process of listing and scoring the failure modes but skip some of the more important steps, such as how to develop your own scoring system and how to identify failure modes effectively. Some commonly used standards and guidelines:

- International Electrotechnical Commission (IEC) standard 812
- Sematech standard E14
- Military standard MIL-STD-1629

16.3 The Basic Steps to an FMEA

For each type of FMEA, the fifteen basic steps are as follows:

1. Determine the Boundaries of the FMEA
2. Gather Documentation and Review Design/Process
3. Build the FMEA Team
4. Determine Customer Usage Profile
5. Develop a Scoring System
6. Brainstorm Potential Failure Modes
7. Transfer the Brainstorming Results to a Spreadsheet
8. List Potential Effects of Each Failure Mode
9. Assign Scores to Each Failure Mode
10. Calculate the Risk Priority Number (RPN)
11. Prioritize the Failure Modes for Action
12. Segregate the RPN Table
13. Take Action to Eliminate/Reduce High-Risk Failure Modes
14. Calculate the Resulting RPN
15. Update the FMEA throughout the Product Life Cycle

16.3.1 Determine the Boundaries of the FMEA

First, determine what type of FMEA you will be performing (see Section 16.1 for the different FMEA types) and the boundaries for the FMEA. If you specifically want to focus on the manufacturing process, make sure you clearly set this as a boundary for your FMEA. One mistake I often see is that engineers try to accomplish too much with a single FMEA resulting in many high-level failure modes. Since there is a lack of focus, they also miss many failure modes.

16.3.2 Gather Documentation and Review Design/Process

Next, review the design or process to which you will be performing the FMEA and gather any documentation that will be helpful during the FMEA process, including block diagrams, interface diagrams, user manuals, schematics, manufacturing procedures, and any other documents you believe will be helpful to the process.

16.3.3 Build the FMEA Team

FMEAs are best performed using a team approach. First, you need to decide who to invite on your team. Make sure you choose a cross-functional group of individuals so that you get a productive exchange of information. This could be representatives from design engineering, software engineering, manufacturing engineering, test engineering, customer service, marketing, quality, and reliability (and possibly other functions as needed). A good size team is six to ten people. If you have fewer, you may not get workable ideas flowing, and if you have more, it may be difficult to control the discussions. Next, you need to choose a facilitator. The facilitator will guide you through the process, make sure that everyone is heard from, and not permit any individual to dominate the conversation. We discuss the role of a facilitator in more detail in Chapter 17.

16.3.4 Determine Customer Usage Profile

As a team, discuss the customer usage profile and reach a consensus on this. This may even be defined in a marketing specification, but make sure the team is clear on what this is before starting the FMEA. I have seen cases where this becomes a hotly debated issue; until this is resolved, you won't have the proper mindset on potential failure modes.

16.3.5 Develop a Scoring System

Developing a scoring system is a very important component of the FMEA process because you need to tailor this scoring system to the specific product you are analyzing in order for the process to yield the proper results. The scores consist of:

- Severity of failure (S)
- Probability of occurrence (P)
- Detection (D)
- Risk priority number (RPN) (multiply S, P, and D together to get RPN).

I recommend using a 1–10 scoring system, but you can use other scales if they work better for you. You don't need to define a score for every value. If you leave any values undefined when creating the scoring system, when you are performing the FMEA, choose the value closest to the situation at hand. You will see examples of not scoring every item in Tables 16.1 through 16.3.

I caution you not to use the scoring system directly out of any of the FMEA standards, such as the ones I listed in Section 16.2. They offer generalized scoring systems, but you need to tailor the scoring systems for your product and your situation for the FMEA to be effective. For example, for the Severity of failure (S) score, if you define a score of 10 for a failure mode that results in death, but the most severe situation that can occur with your product is moderate injury, then your scoring system will be skewed on the low side. At that point, you may decide not to address any failure mode because they all have too low of a score.

16.3.5.1 Severity of Failure (S)

Severity of Failure (S) is a measure of how severe the effects will be if the failure mode does occur. Some companies develop three different scoring tables for severity—the first to address the severity to the equipment, the second to address the severity to the end system, and the third to address the severity to the user.

Table 16.1 shows an example of a severity scoring system we used for a robot that moves semiconductor wafers between pieces of semiconductor processing machines.

Table 16.1: Severity (S) Scoring System Example

	Score	SS=System Failure	SW=Wafer Damage-Level	SH=Human Safety-Impact
Most Severe	10	Catastrophic Failure - Replace Entire System	Multiple Broken Wafers	Death
	9	Failure of a FRU Component, MTTR > 1 Hour	A Single Broken Wafer	Serious Injury
	8	Failure of a FRU Component, MTTR < 1 Hour	Severe Wafer contamination	Minor Injury
	6	Failure that results in reduced throughput	Moderate Wafer Contamination	Possible Injury
	4	Failure that requires a tool reset or recalibration	Slight Wafer Contamination	Possible Injury in Service Mode
	2	Failure that can be corrected during a PM cycle	Unlikely to impact wafers	Unlikely Safety Risk
Least Severe	1	Failure that does not affect system performance	No impact on wafers	No Safety Risk

16.3.5.2 Probability of Occurrence (P)

Probability of Occurrence (P) is how often the failure occurs when creating this scoring system. For this, you have two options. The first option is to use a failure rate table, then assign scores based on the failure rates from your Reliability Prediction. The problem with this approach is that many of the failure modes you uncover may not be component failure modes, and therefore you won't be able to come up with accurate prediction numbers for these failure modes. For example, if you identify a failure mode in which a power supply fails due to a faulty switch, you could assign a failure rate for this based on the failure rate you assigned during your prediction. However, if you identify a failure mode in which a power supply fails because the user plugged a 110VAC power supply into a 220VAC application, you may have a more difficult time determining a failure rate for this occurrence.

The second option is to use more general descriptions for the Probability of Occurrence, such as once per day or once per week. Table 16.2 shows an example of a probability scoring system we used for a medical device company.

Table 16.2: Probability (P) Scoring System Example

	Score	Description
Most Frequent	*10*	Likely to Occur Chronically (Daily or Hourly)
	9	Likely to Occur During One Week of Operation
	8	Likely to Occur During One Month of Operation
	6	Likely to Occur During One Year of Operation
	4	Is Likely to Occur During the Life of the System
	2	A Remote Probability During the Life of the System
Least Frequent	*1*	An Unlikely Probability During the Life of the System

16.3.5.3 Detection (D)

Detection (D) is the ability to detect a failure if it does occur. For Detection, you also have two methods. First, you can develop a scoring system around detecting a failure mode before it occurs. Second, you can develop a scoring system around detecting a failure mode after it occurs so that you can mitigate the failure before the situation gets worse. For example, if your automobile gets low on fuel, the low fuel light will come on, warning you that you will run out of fuel soon. This is the detection method that the automobile manufacturer put in place as a warning so that you can take care of this before the failure mode gets worse (the car stops because you ran out of fuel). Table 16.3 shows an example of a detection scoring system we used for an automobile assembly company. In this example, I have combined the two different methods.

Table 16.3: Detection (D) Scoring System Example

	Score	Description
Most Probable	*1–2*	Very likely it will be detectable before it occurs and after
	3–4	Some ability to detect before it occurs and good ability to detect after
	5–7	No ability to detect before it occurs but good ability to detect after
	8–9	No ability to detect before it occurs but some ability to detect after
Least Probable	10	No ability to detect before it occurs or and some ability to detect after

Note that the table is reversed from the Severity and Probability tables in that a high likelihood event gets a lower score. The reason for this is that Severity and Probability measure a failure occurring, whereas Detection measures a failure being prevented. Also, note that as of this writing of this book, some FMEA guidelines are getting away from using the detection score method because it can be captured as part of the Probability of Occurrence score.

For different types of FMEAs, you will need to come up with scoring systems unique to the FMEA you are performing. For example, if you are performing a Process FMEA, you will need to come up with scoring systems that are unique to failures related to the manufacturing process.

16.3.5.4 Risk Priority Number (RPN)

The RPN is obtained by multiplying the scores together of the Severity, Probability, and Detection. Develop a scoring system with categories that clearly segregate the risks into different categories. The final score is the resulting RPN. For this, develop categories from intolerable to negligible. However, I strongly recommend that you don't assign scores to these yet. You may find that during the FMEA scoring (see Section 16.3.9) the scorers may have a bias towards assigning a lower score to make the failure mode drop down in the RPN category, thus avoiding the need to work on a particular failure mode. I recommend you divide the categories after you have scored all of the failure modes.

Table 16.4: Risk Priority Number (RPN) Scoring System Example

Intolerable Risk	Additional measures are required to ensure adequate safety
Undesirable Risk	Risk is tolerable only if risk reduction is impractical or if reduction costs are grossly disproportionate to the improvement(s) gained. (Requires Executive Mgt. Approval)
Tolerable Risk	The risk is tolerable if the cost of risk reduction will exceed the improvement(s) gained. (Requires Project Mgt. Approval)
Negligible Risk	Acceptable as implemented

16.3.6 Brainstorming Potential Failure Modes

The facilitator must now lead the team through a brainstorming exercise to identify all potential failure modes. There are a number of good brainstorming techniques to help allow ideas to flow freely but also to guide the ideas so that they aren't completely random. The problem with allowing random ideas is that the exercise has no finite bounds, and the team will start getting impatient if there is no end in sight.

Two of my favorite techniques for determining failure modes during a brainstorming session are shown in the Boundary Interface Diagram and the Parameter Diagram.

16.3.6.1 Boundary Interface Diagram

The Boundary Interface Diagram is useful when you have a complex system or your system has interfaces with other systems. Figure 16.1 shows an example of a Boundary Interface Diagram. Here we draw lines showing the interfaces between the different systems. The interfaces can be either physical, energy, material, or data. If you draw the Boundary Interface Diagram on a board during the brainstorming session and identify the different interfaces, then the team can concentrate on the failure modes associated with each interface.

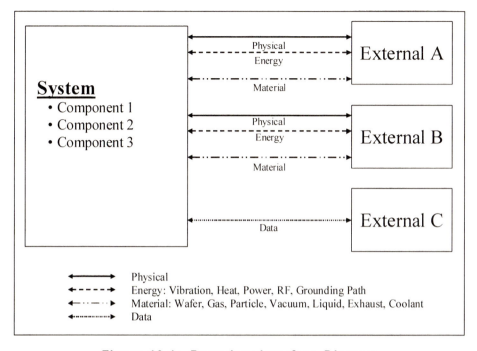

Figure 16.1: Boundary Interface Diagram

16.3.6.2 Parameter Diagram

The Parameter Diagram, or P-Diagram, as it is more commonly called, is a way to focus the brainstorming session into four different areas: Piece to Piece Variations, Environment, Customer Usage/Duty Cycle, and Deterioration. Each area is called a Noise Factor, or a factor that you can't control. Start with one area to determine all of the failure modes pertaining to that area and then move on to the next area. Each failure mode has Inputs, Outputs, Control Factors, and Error States. The definitions of each of the factors:

- Noise Factor—effect of all the uncontrollable factors
- Control Factor—all of the factors that you have the ability to change
- Input—the inputs to the system from another device or from a user (such as a user interfacing with the system)
- Output—what the system is supposed to do
- Error State—these are the resulting effects (this information can feed into Step 8/Section 16.3.8).

Figure 16.2 shows an example of a P-Diagram.

Note the similarity of this technique with Design of Experiments (DOE). With DOE, we also have noise factors, control factors, and errors.

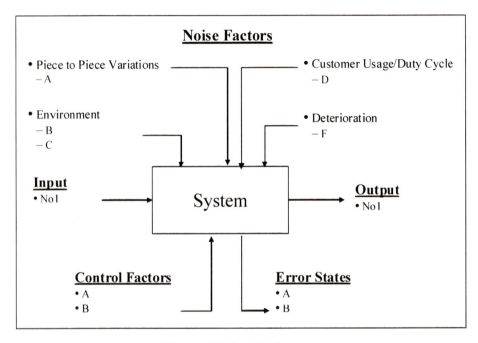

Figure 16.2: P-Diagram

After you complete the brainstorming exercise, you no longer need the entire team for the remaining steps. Many companies make the mistake of involving the entire team in every step of the process. This may result in a very long and tedious process for all of the team members.

16.3.7 Transfer the Brainstorming Results to a Spreadsheet

Now it is time to take the information you gathered in the brainstorming exercise and transfer it into a spreadsheet format. Another common mistake I see teams making is that they use a spreadsheet during the exercise. This tends to derail the brainstorming session because team members will focus instead on filling in the boxes in the spreadsheet and assigning scores.

Table 16.5 shows an example of a spreadsheet you can use. Take note that on the spreadsheet that I have a "u" and "m" next to the columns P, D, and PRN. The "u" is the unmitigated score (the score *before* you take any action), and the "m" is the mitigated score (the score *after* you take action). Note also that I have a column for "Reference to Verification." I recommend you put down the reference showing how you verified the action worked, or if you haven't yet taken the action. This is the reference to the action that will take place when you get to that stage of the product development. For example, if you uncover a failure mode in which the user may wash your product incorrectly, then you will want to call out the proper washing procedure in your user's manual. However, the user's manual has probably not been written yet, so this will be a note to remind you to check to make sure this note gets added to the user's manual when it is written. You can even add another column to the FMEA for "Person Responsible" and assign an action item to the individual responsible for writing the user's manual.

Table 16.5:   FMEA Spreadsheet

Item	Component or Process	Function	Fail Mode or Defect	Local Effect	System Effect	S	Pu	Du	RPNu	Method of Control	Description of Control	Pm	Dm	RPNm	Reference to Verification
1															
2															
3															
4															
5															
6															
7															
8															
9															
10															

For steps 8 and 9 (Sections 16.3.8 and 16.3.9), I am going to describe the vertical approach to the FMEA. Using the vertical approach, we complete a column before moving to the next column rather than completing a row and then moving to the next row (the horizontal approach). As an example, after we have listed all of the failure modes, we will then identify a local effect for each failure mode before identifying any system effects. This is much more efficient than the horizontal approach because you can sort and prioritize as needed to assign effects methodically and score the failure modes.

16.3.8 List Potential Effects of Each Failure Mode

Choose individual members of the team that best understand the product and how failures affect the end system. They should determine the system effects for each failure mode. You may even be able to come up with a table of all possible system effects before starting this exercise so that this process goes much faster and your labeling is more consistent. If your system is complex, you may want to add a column for "Local Effects" as shown in Table 16.5.

16.3.9 Assign Scores to Each Failure Mode

Score each failure mode for severity (S), probability (P), and detection (D). Use the tables you already developed from Section 16.3.5. Once you have identified the failure effects, you should be able to score each failure mode relatively quickly. Often, this only requires one person to score the failure modes.

16.3.10 Calculate the Risk Priority Number (RPN)

After you fill in the S, P, and D values, you should set up formulas in your spreadsheet to calculate the resulting RPN automatically.

16.3.11 Prioritize the Failure Modes for Action

Next, sort the failure modes in the spreadsheet from highest RPN to lowest to determine the order of priority. For any failure modes with a Severity score of 10, you must automatically include these in your list of failure modes to be mitigated regardless of the resulting score.

16.3.12 Segregate the RPN Table

Before starting to develop corrective actions for the failure modes, segregate the RPN table to determine which risks are intolerable, undesirable, tolerable, and negligible. See Table 16.6 for an example of a revised RPN Scoring System with risk scores associated with each risk type. Remember that this is an example; the way you segregate the RPN table may be different.

Table 16.6:   Risk Priority Number (RPN) Scoring System Example

501–1000	**Intolerable Risk**	Additional measures are required to ensure adequate safety
101–500	**Undesirable Risk**	Risk is tolerable only if risk reduction is impractical or if reduction costs are grossly disproportionate to the improvement(s) gained. (Requires Executive Mgt. Approval)
11–100	**Tolerable Risk**	The risk is tolerable if the cost of risk reduction will exceed the improvement(s) gained. (Requires Project Mgt. Approval)
1–10	**Negligible Risk**	Acceptable as implemented

16.3.13 Take Action to Eliminate/Reduce High-Risk Failure Modes

You should now take each failure mode with a high risk number and come up with a method of mitigating or reducing the total risk. Without a method of mitigating each failure mode, the process is only half complete. You may achieve this mitigation by performing one or more of the following:

1. Reduce the probability of the failure occurring.
2. Reduce the severity when the failure does occur.
3. Increase the likelihood of detecting the failure either before the product is released or when the failure occurs in the field.

Note the corrective actions here may require the use of many other techniques described in this book.

16.3.14 Calculate the Resulting RPN

Next, calculate the new RPN based on the actions you took (or the actions you proposed if you haven't yet implemented the actions). Compare against your RPN table to make sure the resulting RPN is now at a tolerable level or negligible level.

16.3.15 Update the FMEA throughout the Product Life Cycle

Your FMEA isn't completed but rather this is as far as you can go in this phase of the product life cycle. As you make changes to your design or as you perform tests, periodically go back and re-visit the FMEA to determine if any new failure modes have appeared. If you redesign a major portion of your product, you should go back to Step 6 (Section 16.3.6) and bring the team together for a brainstorming exercise and then repeat Steps 6–14.

Also, each time you develop corrective actions that you will perform at a future time, go back to your FMEA after you implement the corrective action and confirm that the action was as effective as you predicted it to be when you first developed the mitigated score.

RELIABILITY INTEGRATION: Integrating FMEA with the HALT Plan—Use FMEA to Help Write a HALT Plan
FMEA is a great technique to use prior to writing a HALT Plan because the FMEA can identify how different portions of the product can fail so that you can develop tests to validate the mitigation for these failure modes. See the Inhaler example in Chapter 19. In addition, the FMEA can point out nonrelevant failure modes so that when you are developing your test plan, you choose stresses and levels that avoid finding these failures.

CASE STUDY: FMEA Using a Facilitator
A semiconductor manufacturing equipment company had been using the FMEA technique in the past, but it was relatively ineffective because each time they attempted an FMEA, it resulted in the team arguing over many different details with very little positive results emerging. In addition, various team members commented that the process took too long and they dreaded being part of the process. They asked us to facilitate their next FMEA meeting. First, we bound the FMEA exercise so that we could focus on one specific area. We then took the team through the Boundary Interface Diagram and P-Diagram brainstorming techniques to help the team zero in on specific areas of the design, rather than the random approach they were used to. Once we had a list of the failure modes, we dismissed most of the team and worked with a sub-team to complete the FMEA. This resulted in a much more effective FMEA, uncovering over 50% more relevant failure modes, and we completed it in half the time it took in our client's previous FMEA exercise.

---

17 Facilitation—The Key to a Productive Business Session

Facilitation is the process of designing and running a successful and impartial meeting. With a number of different reliability techniques, it is best to work as a team rather than have one individual perform the task. Examples of these techniques are Goal Setting, FMEAs, Fault Tree Analysis (FTA), DOE, and Software Reliability during design reviews (see Section 31.3.3 for the topic "Facilitation of Team Design Reviews"). When teams get together, often they find that each member of the team has various levels of understanding of the technique being used. Each person was likely taught differently. Therefore, when you get a group of people together in a room it can get quite chaotic; some team members may not even understand the technique. What is needed is a strong facilitator to guide you through the process and make sure that everyone is heard from and no individual dominates the conversation.

17.1 Facilitator Guidelines

For the facilitation to work, you need to put together a team that is willing and able to collaborate. For this to occur, you can use these guidelines:

1. The best size for a team is six to ten people—and each significant technical function should be adequately represented.

2. Consider including individuals from vendor or customer organizations because they have experience with your product from a different viewpoint than people from within your organization.

3. Consider including customer service, field support, marketing, and purchasing because these individuals may be able to provide a unique perspective on particular situations (customer service and field support engineers provide a good customer perspective, while purchasing can provide a good vendor perspective).

4. Not everyone on the team has to be familiar with the product.

5. The team should have at least one product design expert and one man-ufacturing process expert.

6. Each member should have a stake in the process as well as the desire to cooperate towards common goals. This may sound silly because they are being paid by their company to do a job, so of course they should do this. Unfortunately, office politics often will be present, so the facilitator should be looking for team members who have hidden agendas or those who make deals with others (you may need to stop the "you approve my ideas and I will approve yours" situations).

7. Each member of the team must be able to commit a certain amount of time to the project. It does no good to have a key team member present for the first meeting and then be absent for the next. That will disrupt the continuity. On the other hand, there will be team members who aren't needed for every meeting. Perhaps they are only needed for the kickoff meeting to provide a high level perspective. Other team members may be critical and are needed for the brainstorming sessions but may not be needed for some of the scoring/metric setting meetings. For example, with FMEA, once you list all of the failure modes, it is time to score them. At the beginning of the project, the team should decide on a scoring sys-tem that will work for the project. Therefore, the entire team isn't needed to score each individual failure mode as long as the scoring system de-veloped at the beginning was clear.

8. Members should actively participate, listen, and voice any disagreements in a constructive manner. They should be open to ideas from other mem-bers but also be prepared to contribute to the conversation. They should actively listen to others and respect alternative views. Remember the 1957 Sidney Lumet movie *12 Angry Men*? In this film, a single dissenting juror in a murder trial slowly manages to convince the other jurors that the case isn't as obviously clear as it seemed in court. The interesting part is that many of the other jurors were happy to go along with the majority sim-ply because they had other plans and wanted to get it over. We've seen this frequently happen in meetings in which members of the team will go along with anything the majority says just to hurry the conversation. The facilitator can't let this happen.

17.2 The Facilitator's Responsibilities

The roles and responsibilities of the facilitator are:

1. Set up and lead the meetings.

2. Ensure the team has the necessary resources. Resources can be per-sonnel, time, and conference rooms. If some members are calling in to the meeting, make sure you have a clear conference phone (video con-

ference capability and file sharing capability via a good web-conference tool is even better).

3. Make sure no one brings a laptop (unless it is needed for the meeting) and make sure all mobile phones are switched off. The brainstorming exercise isn't a time to allow multi-tasking.

4. Make sure the team is making progress throughout the process.

5. Make sure the group isn't caught up in one particular issue too long. This is probably one of the trickiest areas because you don't want to cut off healthy conversations, but you also don't want to go too far down one path and risk losing the interest of all of the other members, or get too far off the subject. The facilitator should know how to tell the group that the current conversation is healthy and meaningful but needs to be taken offline.

6. Break up the meetings into several shorter meetings. Don't try to meet for more than two hours at a time—longer than this and the team members will lose interest. I've seen many FMEA processes fail because members experience that the process takes too long with very little benefit.

7. Make sure ideas are free-flowing and not cut off.

8. Make sure the team listens to all of the members of the team that are providing input.

9. Make sure the team reaches a consensus. For a team to reach a consensus the following should take place:

    a. There should be 100% participation among the members of the team.

    b. The requirement isn't 100% agreement, but you should have 100% commitment to the process. Unlike a jury that has to deliver a unanimous decision, it isn't necessary for every decision in an FMEA to be unanimous.

    c. The majority doesn't rule. I've seen many technical discussions end by taking a vote, and this is absurd. Shouldn't technical discussions be decided with facts, not votes? Some types of decisions, such as "Do you want to add that feature to the product?" can be decided based on a majority vote. Decisions such as "Should you fix that potential safety hazard?" can't be decided based on a majority vote. These types of decisions should be decided using facts.

10. Recognize when you reach a consensus. For this to occur, the team and its members should answer yes to the following:

    a. Have I honestly listened?

    b. Have I been heard and understood?

    c. Will I support the decision?

    d. Will I say "We decided," as opposed to "I told them and they followed my recommendation?"

11. Handle difficult people. Difficult people can be:

   a. The individual who talks too much.

   b. The individual who talks too little or is disengaged (reads emails or leaves the room to answer the phone).

   c. The individual who continues to go off topic.

   d. Aggressive people—shouting to get their point across, or worse, ridiculing other peoples' ideas.

   e. Passive-aggressive people—these folks are in constant disagreement but instead of outwardly disagreeing, they close up and say nothing. This is different from the individual who talks too little and is multi-tasking. Passive-aggressive people may not be multitasking and may "appear" engaged, nodding their heads and voting along with the group, when in fact they disagree but are afraid of confrontation.

17.3 Who Can Be a Facilitator?

Any member of the team can facilitate so long as they follow the facilitation guidelines. Often, the facilitator is a member of the quality or reliability staff. External consultants can be great facilitators as they can avoid the internal office politics. External consultants also can help select a facilitator from within the company and train that individual.

RELIABILITY INTEGRATION: Integrating Facilitation with Reliability Techniques—Facilitation Isn't Just for FMEAs
When people think of facilitating reliability discussions, they often think of FMEAs. Facilitation is a natural technique to use during the FMEA process, but it can also be used as a valuable technique for many other reliability techniques, such as Goal Setting, FTA, and DOE, and in Software Reliability during design reviews.

CASE STUDY: Facilitating a Goal Setting Meeting
A test equipment company asked us to facilitate their discussions because they had never conducted a Goal Setting meeting previously. The first step we took was to review the reliability data from the previous design and identify the changes being made from that design to this new design. When we first asked marketing to list the differences between the product versions, they said that the product was virtually identical, except for a few new features. Then, when we went through the system architecture, we found that the following components had changed: CPU speed and architecture (from single to dual processor), memory size and density, storage type (from disk to solid state), cooling design, power supply load and amount of derating on the power supply, and the software. By facilitating this meeting, we learned that in reality their product had changed quite a bit. In addition, they wanted to improve their reliability "significantly" over the previous design, but they didn't have much data to show what the actual field reliability was for the previous design. We sent a team off to review the test and field data from the previous design. Once we had this data, we reconvened and came up with a goal statement of how much they wanted to improve on the previous design. We then determined the contribution of failures from the previous design, as well as the first pass estimates on what we could expect from each new assembly, given the changes. This gave us much more realistic goals. At the end of the program, we measured the field data compared to the reliability goals and showed we were able to achieve their goals.

18 A Good Kind of Fault—FTA

Fault Tree Analysis (FTA) is a top-down approach to failure mode analysis. You can use an FTA to identify failures and to eliminate the cause of the failure. An FTA is a systematic, deductive method for defining a single specific undesirable event and determining all possible failures that could cause the event in question to occur.

An FTA can be both very useful in the initial product design phase as an evaluation technique, and it can also be used as a powerful troubleshooting technique after an event (or proposed event) has taken place.

FTAs and FMEAs are very similar in this regard, but the goal is much different. With an FMEA, you try to identify all possible failure modes in a system and the effects of these failure modes. With an FTA, you start with one specific failure effect and then identify only those failure modes that have the highest probability of causing the particular effect.

When you perform an FTA, start with an undesired event. The undesired event constitutes the top event in an FTA diagram. Then conduct a brainstorming session to determine the possible failure modes that can result in this undesired effect.

Figure 18.1 shows an FTA diagram for a computer with the end effect of "Computer Fails." In this case, you know the end effect, and you can draw an FTA to determine the possible cause of the failure.

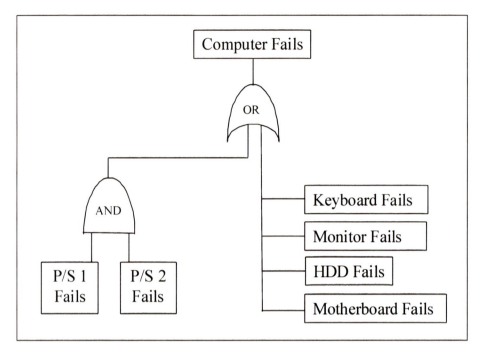

Figure 18.1: FTA for a Computer

18.1 How to Decide Whether to Use an FTA or an FMEA

FTA is preferred over FMEA when:

- You can identify a small number of top events.
- Product functionality is highly complex.
- The product isn't repairable once initiated.

FMEA is preferred over FTA when:

- The events can't be limited to a small number.
- Multiple successful functional profiles are possible.
- Identification of "all possible" failure modes is important.

18.2 FTA During Reliability Apportionment

Another use for FTA is during Reliability Apportionment. Figure 18.2 shows the same computer example as in Figure 18.1, but it is performed in conjunction with Reliability Apportionment. Notice that instead of analyzing for failure situations, we are apportioning for success. We can then assign reliability numbers to each block, as we did in Figures 10.1 and 10.2.

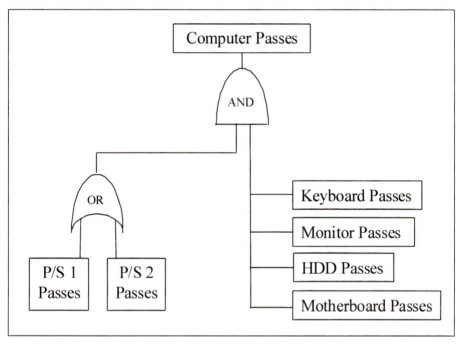

Figure 18.2: FTA using Reliability Apportionment

RELIABILITY INTEGRATION: Integrating FTA with HALT—Using FTA during HALT Planning
When an FMEA identifies a critical effect, FTA is often deployed to evaluate all possible failure modes that can also cause the same critical effect. This is especially helpful when planning a Highly Accelerated Life Test (HALT), so that you apply the appropriate stresses and can easily troubleshoot the failure if you expose the critical effect during HALT.

CASE STUDY: Using FTAs During Failure Analysis
Airplane crash investigators use FTA during crash investigations. First, they consult the FMEA to determine the potential ways the airplane can crash. They then identify the different failure modes that could cause the event. From that, they review the evidence (such as the black box recorder, crash site investigation data, weather reports, and material analysis) and assign probabilities of occurrence for each possible scenario. This aids them in their investigation to determine which failure analysis path to follow.

19 FMEA Should Be Used to Drive Testing

The need for better reliability test programs is being driven by:

1. Industries are competing on reliability more and more.
2. Companies need to develop more reliable products, faster.

Most engineers write reliability test plans generically, or they blindly follow industry standards. This results in too little testing. You should tailor test plans to fit customer use profiles. In order to write better test plans, you must first understand the use environment and the key risks to the design.

The best technique for this is **FMEA**. See Chapter 16 for more information on FMEAs. Once you have identified and prioritized the risks, it is time to develop mitigating methods. Often, the best mitigating methods are with reliability testing. Stated another way, you can't know what to test for unless you understand the key risks. Therefore, FMEA is one of the best sources of input for your Reliability Test Plan (RTP).

The best way to illustrate this technique is with an example. I have included here an example of a medical inhalation product we worked on for a client. The device consisted of a non-rechargeable battery, a dosing dial, electronics, and a disposable vial of medicine, all in a plastic enclosure. The heart of this product was a one-inch diameter metal disk that had about one million perforated holes. This piece of metal contributed significantly to the overall product cost, so a failure of this component would result in discarding the entire product. See Figure 19.1 for pictures of the device.

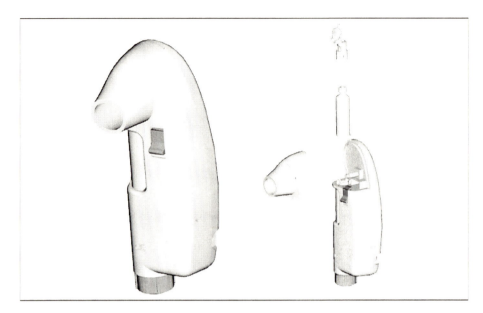

Figure 19.1: Picture of Medical Inhalation Device (left) and Breakout View (right)

What types of tests can you think of for this product? We polled a group of people and asked them what tests they thought were appropriate and this is what they came up with:

1. High/low temperature test
2. Temperature cycling test
3. Vibration test
4. Drop test
5. Shock test
6. Crush test
7. Humidity test
8. Altitude test

   Did they miss any?

   Then we performed FMEA and came up with the following:

9. Different Cleaning Solutions: For this, the user's manual stated to use only soap and water, not to use alcohol, and not to put the product in the dishwasher. During the FMEA, we identified a failure mode "What would happen if the user doesn't follow the cleaning instruction?" Would it fail and if so, how? We decided to set up a test to clean with alcohol and

guess what? The housing started cracking. Our recommendation was then to change the material because there will likely be users that don't read the manual or ignore the instructions in the manual and use alcohol.

(I've only listed here a partial subset of the tests we performed as a result of the FMEA.)

FIRST RULE OF USER'S MANUALS: Users won't read them. Don't put any information in a user's manual that you expect and require someone to read in order to determine the functionality of the product or worse, rely on the manual to point out a safety hazard because most users won't read the manual. Make everything intuitive, anticipate what the user will and won't do, and design for that.

10. Poking Test: Due to the sensitive nature of the perforated disk, we were worried that contact with a sharp object could puncture the disk and either cause the product to leak or, even worse, cause an overdose situation. There are many different scenarios that could result in this. The most likely scenario would be if the user put the product in their pocket or purse where it would subsequently come in contact with sharp objects, such as a pen, a golf tee, keys, or perhaps even a screwdriver.

11. Clogging Test: As with the poking test, if the product comes into contact with a substance that could clog the holes, the user could get an underdose situation. The most likely scenario would be if the user put the product in a pocket or purse where it would subsequently come into contact with an open lipstick container or a leaking pen. We even considered a situation where the product gets put on the ground or seat of a car near motor oil or some other type of sticky substance, such as candy. The user could have motor oil or other chemical on their hands, or grease from a hamburger they just ate. Can the plastic withstand these types of chemicals? Could these possibly come in contact with the metal disk and clog it? These are areas you need to consider during the product design, especially during the testing.

12. Cap Tether Test: To solve the two previous situations, we considered adding a cap to the product, but caps get lost very easily, so we designed the cap with a tether. When using a tethered cap, the next logical test was to test the tether mechanism to ensure it wasn't only strong and able to withstand many thousands of cycles, but that it was ergonomically and aesthetically pleasing so that the user doesn't purposely take the cap and tether off the product.

13. Button Life: We determined that it is possible that the buttons could wear out, given the use profile for the product and the buttons. Therefore, we set up a separate button life test and exercised the buttons using normal force to simulate an average user, as well as abnormally high force to simulate an exceptionally strong user.

14. Battery Insertion Test: Is it possible to put the wrong type of batteries in the product and damage it? Products that take rechargeable batteries often won't run on disposable batteries, and vice versa.

15. Battery Life Test: This product ran on batteries, so what would happen when the batteries start running low? Is it possible for the product to begin providing the dose and then stop in the middle if the batteries run out? Does the user know if the dose was administered? Is it possible that the user got no dosage? Or, just as bad, is it possible that the user got a dosage but didn't know it (so when the battery is replaced, the device reapplied the dosage)? We recommended a battery life test to cover these situations.

As you can see by this example, we would have missed many of the potential failure modes had we not used FMEA to help write our Reliability Test Plan (RTP). In order to develop more meaningful test plans and find really relevant reliability issues, you should use the FMEA process.

RELIABILITY INTEGRATION: Integrating FMEA When Writing an RTP
We have described in detail in this chapter how you can use an FMEA when writing an RTP. An FMEA will identify and prioritize the risks in your product, and the RTP will then put forth test methods to mitigate these risks.

CASE STUDY: Test Plan for a Battery
We were working with a medical infusion pump company developing a test plan for their product. When we reviewed the battery design, we looked at the usage profile for the battery and determined that the product ran mostly on power when the patient was in the hospital room. The battery was used only one to two hours per day when the patient was being moved between hospital rooms. This told us that the battery use profile was made up of mostly shallow discharge cycles (as opposed to deep discharge cycles that are more prevalent with mobile phones). We reviewed the battery manufacturer's life test data and determined that they had performed both types of life tests-shallow discharge as well as deep discharge. We reviewed their life test data based on shallow discharging and found the test data to be acceptable and met our projected life test goals. This saved us a lot of time and money (since we didn't have to perform our own life testing), while still meeting our reliability goals.

20 Reliability During Component Selection

Component Selection typically falls within the responsibility of component engineering, but there are significant reliability implications to this process. Therefore, I strongly recommend that the reliability group be involved in this process and set some guidelines to follow. Components today are significantly more reliable than components from even five to ten years ago. However, most designs have a few components that account for the majority of reliability or quality issues. It is these few components that you should analyze in more detail during the design to mitigate these issues before they occur.

The purpose of selecting the appropriate components for the particular application and environment is to optimize cost versus reliability (maximizing reliability at the lowest possible cost) in the given use environment. See Section 3.1 for more information on optimizing cost versus reliability. When selecting components, you first should understand the environment that the end-product is going into and the overall reliability goals. Then, choose the appropriate grade of components (commercial, industrial, or military). You may even need to specify additional component screening, especially if you are forced to use a lower grade component than you desire. This could either be because the higher grade component isn't available or is too expensive. Many military suppliers are using commercial off-the-shelf (COTS) components for these reasons. They often must specify additional screening because the COTS component specifications are lower than the specifications for their system. In addition, check to ensure that the component manufacturer is reputable with a proven track record.

Four common parameters that you should check:

1. Component qualification parameters
2. Manufacturing stress parameters
3. Ongoing reliability tests for the product

4. Manufacturer qualification (e.g., military certified, ISO certified)

As part of the component selection process, you should develop a reliability critical item list. This is a list of components that require special attention; therefore, reduction of this list is a key goal early in a program. You should put components on this list for a variety of reasons:

1. Low reliability
2. High criticality
3. Long-lead time
4. Approaching obsolescence (see the Reliability Integration section in this chapter)
5. Risk of receiving counterfeit components (see Note 1)
6. High cost
7. Improper derating
8. Possible tolerance issues between components
9. Component requires uprating
10. Component parameters aren't guaranteed by manufacturer (see Note 2)

NOTE 1: Component Counterfeiting is becoming a bigger problem every year. China is one of the main countries responsible for this, but just about every country, including the United States, is guilty of counterfeiting. In fact, the electronics industry isn't the only industry being affected. The clothing industry and the wine industry are two other industries that are also being hit very hard by this growing issue.

NOTE 2: When the situation arises in which your supplier can't or won't guarantee component parameters, then you should write a custom specification for the product and use a process we call Component Parameter Testing. See the case study at the end of this chapter for an example of Component Parameter Testing. It is best to perform this testing when inspecting incoming components. However, if you have many components that fall into this category, it is worth finding a third-party vendor to perform this testing for you.

RELIABILITY INTEGRATION: Integrating Electronic Design Automation (EDA) Techniques When Facing Obsolete Components

We are facing the issue of component obsolescence now more than ever before. Therefore, component engineering tends to monitor a design well beyond release all the way to its end of life. Your component engineers must find replacement parts when parts go obsolete. Additionally, as components become more complex, simply comparing specification sheets won't work anymore. You need to order the replacement part, put it into the circuit, and perform qualification testing to ensure nothing subtle changed within the design. You should then perform HALT to make sure the product has as much margin with the new component as with the previous component. This can be a costly and time-consuming process if it turns out that the component isn't a good match. There is a better way to do a cursory check to make sure the new component meets all of the basic parameters—by using Electronic Design Automation (EDA). The same technique that was used by your designers to design the circuit can be used by your component engineers during component obsolescence evaluation. It won't guarantee the product will work-testing should still be performed—but it will tell you quickly and inexpensively if the component *won't* work.

CASE STUDY: Component Parameter Testing

We were working with a company making controllers for an outdoor application. They were using a TRIAC on their design (a TRIAC is a component that controls and conducts current flow during both alterations of an alternating current (AC) cycle). They found a TRIAC from a second source vendor; all of the specifications appeared to match up well, but when the company put the component in the application, it started failing. Why did this happen? Because when the design engineer chose the original TRIAC, he knew something about the TRIAC's performance that didn't appear on the datasheet. He didn't document this, create a special specification sheet, or ask the vendor to add this information to their datasheet. He then left the company, and the new design engineer had no knowledge of any of this. When the new engineer chose the second source component, it didn't have the same performance characteristics needed for the undocumented parameter.

We reviewed the specification sheets more closely and determined which parameter was causing the problem. We discussed this with the second source TRIAC manufacturer, and they told us they couldn't control the parameter as tightly as we needed. Therefore, we recommended to our client to disqualify this manufacturer. We found two other manufacturers that could control this parameter as tightly as we needed for the design, and we recommended to our client to purchase samples from these two manufacturers for further testing to confirm they would work as reliably as the original component.

In working with our clients, we have found that about five to ten percent of the active components will have a parameter that the design engineer requires for operation but isn't guaranteed by the component manufacturer. Component Parameter Testing can reduce the probability of failures with these components.

21 Predictions—Better Than a Crystal Ball

A **Reliability Prediction** is a method of calculating the reliability of a product or piece of a product by assigning a failure rate to each individual component, then summing all of the failure rates.

Reliability Predictions are used during the component selection phase to:

1. Help assess the effect of product reliability on the quantity of spare systems required, which feeds into the life cycle cost (LCC) model.
2. Provide input to the Warranty Analysis and help with Warranty Projections.
3. Assist in deciding which product to purchase from a list of competing products by comparing failure rate numbers.
4. Set achievable in-service performance standards against which to judge actual performance and stimulate action.
5. Help make initial assessments of complex systems.
6. Understand how often different components of the system are likely to fail, even for redundant components.

21.1 History of the Reliability Prediction

In the 1950s, the Department of Defense (DOD) first standardized electronics Reliability Predictions through the analysis of historical data. This led to the publication of the first edition of the military standard MIL-HDBK-217 in 1961, providing the basis of Reliability Predictions that is still widely used today.

In 1994, U.S. Secretary of Defense William Perry published his pivotal memorandum titled "Specifications & Standards—A New Way of Doing Business." One consequence of the memo was that military acquisition changed dramatically, and the DOD cancelled many military standards in favor of commercial standards and practices. Another consequence of the memo was that the DOD stopped updating MIL-HDBK-217 and instead looked to industry organizations to provide updated Reliability Prediction methods.

21.2 Should Predictions Be a Qualitative or Quantitative Technique?

The Reliability Prediction is one of the most controversial topics in the area of reliability. The controversy centers on whether or not one can accurately predict field reliability. Most engineers agree that traditional Reliability Prediction methods using the old MIL-HDBK-217 and even the newer Telcordia SR332 aren't very accurate. However, some engineers continue to use these methods because they realize the value in a prediction lies beyond the end mean time between failure (MTBF) number. These engineers use predictions to help influence design decisions because predictions can point out design weaknesses and are great for trade-off analyses. We discuss this more in Section 21.4.

Other engineers are looking for ways to increase the accuracy of predictions. These engineers are turning to more progressive methods of performing predictions such as:

1. PRISM®[1] is the System Reliability Center (SRC) software tool that ties together several tools into a comprehensive system Reliability Prediction methodology. The PRISM® concept accounts for the myriad of factors that can influence system reliability, combining all those factors into an integrated system Reliability Assessment resource. Some of the factors and areas PRISM® predictions consider that MIL-HDBK-217 and Telcordia SR332 don't are process grading factors, ability to use predecessor system test and field data, and non-operating failure rate models. One drawback of using PRISM® is that it could take significantly longer to perform a PRISM® prediction because of the time it takes to determine all of these extra factors. If you are uncertain of any of the factors, or you guess incorrectly for any of the factors, your prediction may not be more accurate than if you had used MIL-HDBK-217 or Telcordia SR332. In addition, PRISM®, like most other prediction methods, starts with a base prediction number, and this number is typically from outdated information.

2. 217Plus™[2] is the latest Reliability Prediction methodology available from the DOD's Center of Excellence in reliability and is intended to replace MIL-HDBK-217. It was developed by the Reliability Information Analysis

---

1. PRISM® is a registered mark of the System Reliability Center (SRC).
2. 217Plus™ is a trademark of RiAC.

Center (RiAC), funded by the DOD, and sponsored by the Defense Technical Information Center (DTIC). 217Plus™ is based on the electronic failure rate data contained in the RiAC databases as of September 2005. Because the RiAC is funded to update this data continuously, users of 217Plus™ will benefit from technology updates and current failure rate experiences. 217Plus™ predictions include many of the same factors and areas that PRISM® covers (that MIL-HDBK-217 and Telcordia SR332 don't) and include a few additional ones such as models for more types of components. The same drawback exists for 217Plus™ as with PRISM®. 217Plus™ predictions take about the same amount of time as PRISM® predictions.

3. MIL217G is an effort being developed by the VITA 51 Working Group. The VITA 51 Working Group was formed in 2004 to investigate the state of the Reliability Prediction industry and develop a method to address electronics failure rate prediction issues. They found that the MIL-HDBK-217 method had become obsolete compared with current electronics technologies; however, it remained the most common method used in industry to predict electronics reliability. Many companies have developed proprietary and unique adjustment factors to MIL-HDBK-217 for their Reliability Predictions, leading to unsubstantiated reliability analyses. This poses a problem for system integrators and customers, who no longer have credible or consistent predictions from the supply chain. As of the writing of this book, MIL217G isn't currently completed or being used commercially, so we can't assess the effectiveness of this method.

4. Physics of Failure (POF) predictions. A few organizations have developed POF models for many different components and component failure mechanisms, and they are proving that these models are more accurate than the old prediction methods. One downside to POF models is that they take significantly longer to predict per component, so we recommend you use POF models only on the components on your reliability critical item list. See Chapter 24 for more information on POF.

21.3 Different Types of Predictions

Here are six different types of Reliability Predictions you can perform:

1. Parts Count Analysis
2. Parts Stress Analysis
3. End-of-Life (EOL) Analysis
4. Service-Affecting Analysis
5. Availability Analysis
6. Field Reliability Analysis

21.3.1 Parts Count Analysis

Parts Count Analysis estimates a product's overall reliability using industry standard failure rates for each component. Telcordia SR332 is commonly used for commercial products, and MIL-HDBK-217 is commonly used for defense products. Section 21.2 describes other methods you can use. You typically perform this analysis at the assembly level but you can also perform it at the system level.

21.3.2 Parts Stress Analysis

Parts Stress Analysis is performed after a basic analysis for a more refined overall MTBF. It considers the influence of additional factors on each component (quality level of components, temperature, electrical stress, environment, and first year reliability estimates based on proposed manufacturing screening). As with the parts count analysis method, Telcordia SR332 and MIL-HDBK-217 are commonly used, but Section 21.2 describes other methods you can use. You typically perform this analysis at the assembly level as well, but you can also perform it at the system level.

21.3.3 End-of-Life (EOL) Analysis

End-of-Life (EOL) Analysis estimates the onset of wear-out failures for consumables (e.g., electrolytic capacitors, fans, motors, drives). This can aid in planning in-warranty costs and scheduling preventive maintenance. This analysis is typically performed on components within an assembly or system. The POF methodology works well for performing an EOL Analysis. See Chapter 28 for more information on EOL Analysis.

21.3.4 Service-Affecting Analysis

Service-Affecting Analysis can be performed after the Parts Count Analysis. Rather than considering every component failure, it discounts failures that don't affect system performance. Acceptable minimum level of performance is carefully defined and considered across the product. This analysis can be performed at the assembly level or system level.

21.3.5 Availability Analysis

Availability Analysis estimates uptime by considering failures that take down an entire system. You should model the architecture, with redundancies, as a combination of series and parallel elements. You can use a variety of techniques to model redundancies. The Markov Model is one of the most popular methods. Telcordia standard SR1171 "Methods and Procedures for System Reliability Analysis" offers guidelines around different types of modeling methods. You typically perform this analysis at the system level. See Section 27.3 for more information on Availability.

21.3.6 Field Reliability Analysis

Field Reliability Analysis collects data from the field database and uses analytical techniques, such as Weibull Analysis, to determine not only the current reliability figures, but also where in the reliability life cycle your product currently is (see Figures 9.1 and 9.2 for examples of the Reliability "Bathtub" Curve). This will tell you whether the product has a decreasing failure rate, constant failure rate, or increasing failure rate. You can perform this analysis at the component, assembly, or system level.

21.4 Predictions Beyond MTBF Numbers

For most of this chapter, we discussed Reliability Predictions as a method to estimate MTBF. However, predictions can serve much more than this. Predictions can also:

1. Drive design trade-off analyses. For example, you can compare a design that contains many simple components to another design that contains fewer but more complex components. The system with fewer components is usually more reliable, but not always—it may depend on the level of complexity for the components in each design.
2. Make early decisions as to the need for redundancy. Even though your prediction may not be accurate, it will be close enough in determining if you will need to employ redundancy to meet your reliability goals.
3. As part of a prediction, you can perform a stress analysis and thermal analysis to determine the stress factor on each component. When you do this, you will learn valuable information about which component types are most sensitive to electrical stress and thermal stress. This can be useful during testing so you know what electrical stresses and thermal stresses to apply and which components to monitor, because they have a higher probability of failing.
4. Predictions can be used as input to your FMEA. In an FMEA, you will need to determine the probability of occurrence of each failure mode. You can use the results of the prediction to supply this information. Even if the failure rates aren't accurate, our experience shows that they are accurate relative to each other. In other words, if you predict that a specific component will fail more often than another component, this will likely be true even if you are off by an order of magnitude on each. For an FMEA, you are scoring failure modes relative to one another to determine which failure modes to address.
5. During the prediction, you will need to research exotic components on which you can't find information in handbooks and databases because the component technology is too new. The information you obtain during this research can be used as input to many other areas of your reliability program.

6. Compare the prediction results from previous predictions compared with the field results after you deploy the product. From this, create a prediction-to-field factor that you can apply to future predictions.

RELIABILITY INTEGRATION: Integrating Predictions with Reliability Goals

You can use Reliability Predictions as an early indicator of whether a design has the potential of meeting your reliability goals. Using predictions, you can calculate a product's reliability level with a reasonable level of accuracy. From this, you can then determine how much re-design and how much testing you will need to ensure that you can meet your reliability goals. If your predictions are below your goals, perform design trade-off analyses, such as adding fans or heat sinks, to determine the effects of each change. If your predictions are above your goals, then determine how much testing you will need to perform to confirm these margins.

CASE STUDY: Reliability Prediction used with Thermal Analysis

Our client manufactured circuit boards and sold them to system integrators. They needed to determine if it was better to use a heat sink on a CPU or a CPU cooling fan. A cooling fan for the entire system wasn't an option because our client only supplied the board to his customers and our client didn't have any control over system-level cooling decisions. The thermal analysis showed that the heat sink could cool the CPU by 10 C, and the fan could cool the CPU by 20 C. The CPU was near the failure point that we discovered during HALT, and therefore this 10 C increase was critical. The failure rate of the fan increased the overall failure rate of the board by about 10%. The increase in temperature of 10 C on the CPU increased the CPU failure rate by 50%. Therefore the net failure rate was significantly higher with the heat sink than with the fan. The cost of each failure was equivalent because a failure in the field would require replacement of the entire board. The cost of the fan increased the overall cost of the board by 5% more than the heat sink, but the increase in failure rate using the heat sink would have increased the total cost by about 20% (the total cost includes the cost of field failures as well as production costs), so we recommended that our client migrate to the fan solution. Table 21.1 shows the trade-off study in more detail.

Table 21.1:   Trade-Off between Fan and Heat Sink

	$\triangle T$	$\triangle FR$	$\triangle$ Cost of Parts	$\triangle$ Total Cost
Heat Sink	10°C	50%	0%	20%
Fan	20°C	10%	5%	0%

22 Thermal Analysis and Reliability, the Hottest Couple in Town

Thermal Analysis is used to estimate the temperature distribution throughout a product based on the Thermal Boundary Conditions and specified heat sources. Thermal Analysis is closely tied to reliability through models like the well known Arrhenius Model, which describes how component reliability is adversely affected as the temperature increases. You should manage heat generation because it is impossible to eliminate all sources of heat. This may involve performing trade-off analyses to help you identify how much the increased heat will affect the reliability of your product. From this, you will be able to determine if the resultant component reliability is acceptable. If you can't meet your reliability targets, you still have a few options, such as reducing the heat using a different method of cooling, moving the components to another area of the circuit board or the system, purchasing higher grade components, or adding redundancy.

Engineering teams typically don't understand thermal risks until after they build and test the first prototype. Unfortunately, this comes too late in the design cycle because the cooling solution is often tied very closely with the mechanical architecture of a product. It may be possible to make minor changes such as increasing the speed of a fan, but it may not be possible to make major changes like adding redundancy or moving components around the system. Understanding and mitigating thermal issues prior to the completion of the design will save a lot of time and money.

If you are going to quantify the effects of temperature on product reliability, then you will need to make sure you choose the proper acceleration model. The Arrhenius Model may not be applicable—it all depends on the failure mechanisms of the components in question. Even if the Arrhenius Model is applicable, you will then need to determine the activation energy of the failure mechanism. The activation energy is the minimum energy required to start a chemical reaction, which is the process that occurs when a substance changes into another substance.

Thermal analysis will allow your design engineers to model the thermal mass of the product or portions of the product and review the temperature distribution, thermal gradient, and thermal flux. They use this information to help make informed decisions about how to modify the architecture of the product as early as possible in the design cycle.

Thermal loads are also important sources of mechanical loading. Even with a uniform temperature applied, an assembly of components having materials with different thermal expansion coefficients can experience significant stress. Cyclic thermal loads can lead directly to fatigue failures due to the cyclic stresses that are generated. You can use Finite Element Analysis (FEA) for either steady state or transient (time-dependent) heat transfer. You can predict the temperature distributions throughout an assembly and the areas of high thermal gradients. You can combine the thermal field with the mechanical properties of the components in order to predict stress and strain fields associated with the thermal loading.

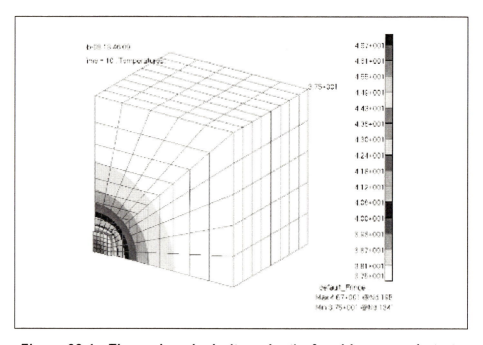

Figure 22.1: Thermal analysis (transient) of a chip on a substrate using the MSC.Nastran finite element software. The temperature distribution is shown as a function of time and position.[3]

3. MSC.Nastran is from MSC Software Corporation

RELIABILITY INTEGRATION: Integrating Thermal Analysis with HALT—Thermal Analysis Can Help in the HALT Planning Process
Proper thermal analysis during the design phase can pinpoint thermal issues, which you can then verify when performing HALT. The Reliability Prediction will reveal which components' failure rates are most affected by temperature. Once you identify these components, apply thermocouples to them during HALT and monitor them closely during testing because they have a higher probability of failing.

CASE STUDY: Thermal Analysis Used to Help Plan a HALT
For a medical device company, we performed a thermal analysis during the design phase. As a result of this analysis, we discovered a few heat-generating areas of the design that were affecting some of the sensitive fiber optics. We decided to run this configuration through HALT and monitor this area of the product using thermocouples while carefully measuring the bit error rate for any degradation in performance of the fiber optics. We found that under ambient conditions, the fiber optics were safe, but if the environment reached its specified maximum, the life of the fiber optics would degrade rapidly. Our client relocated the fiber optics closer to the fans to reduce their temperature. This reduced the temperature on the fiber optics by 10°C, and we calculated that this would prolong their life by several years.

23 If You Want Your Design to Rate, You'd Better Derate

Derating is the practice of operating at a lower stress condition than the rating specified for a component. Derating a component reduces its failure rate. The more you derate a component, the lower its failure rate becomes. The amount the derating affects the failure rate is technology-dependent. Using this process, you can ensure that you are using the most cost-effective components with the appropriate reliability goals in mind. The benefits of performing a Derating Analysis are to optimize the component size and cost with the level of reliability. As a rule of thumb, as you derate, the reliability increases, but the component size and the component cost also increase.

As part of derating, you should select the appropriate components for the particular application and environment. The demand for reliability continues to increase at the same time that management continues to drive costs down. Therefore, there is an increasing need for proper derating and proper component selection to match end-use environments. There are numerous derating guidelines already in existence. None will exactly meet your needs, but some will be closer than others. It is important to choose the guideline that most closely matches your products and the stresses they will be subjected to (both electrical and environmental stresses). After this, you should tailor the guideline to your specific application. Examples of guidelines available in the industry:

- Rome Air Development Center's standard RADC-TR-84-254, "Reliability Derating Procedures"

- Jet Propulsion Laboratory standard JPL-D-8545, "Derating Guidelines"

- Naval Sea Systems Command (NAVSEA) SD-18, "Part Requirement and Application Guide"

- NAVSEA TE000-AB-GTP-010, "Parts Derating Requirements and Application Manual for Navy Electronic Equipment"

- Military standard MIL-STD-975, "NASA Standard Electrical Parts List," Appendix A

- Military handbook MIL-HDBK-338, "Electronic Design Handbook"

- Military standard MIL-STD-1547, "Electronic Parts, Materials, and Processes for Space and Launch Vehicles"

- Association Connecting Electronics Industries' standard IPC-9592 Section, "Requirements for Power Conversion Devices for the Computer and Tele-communications Industries," Appendix A, "Derating Guidelines"

- Telcordia SR332, Issue 2, "Reliability Prediction Procedure for Electronic Equipment," Section 9–2 "Electrical Stress Factor"

It is likely that you are using certain components in a unique way and may require the derating of these components to be different than what is stated in the derating guideline you chose. Your design engineers should be part of the development of the guidelines to ensure they adopt the guidelines. By applying these guidelines, you will determine if there are any components that are over-stressed according to the guidelines you have defined. You can then feed this information directly into the Reliability Prediction model to determine the effect that the overstress has on each component's reliability. Remember, though, that you are simply working to a guideline—if you exceed a guideline limit on one component, you may still be able to compensate in other areas where component stresses are below the guideline limits.

Another technique that is closely related to derating is uprating. This is basically the opposite of derating because, with uprating, you need to use a component outside of its specifications, and you need to determine how much of a reliability impact this will have on your design. One industry that practices uprating all the time is the oil industry. Oil drills are drilling down as deep as seven miles into the earth; at those depths, temperatures are approaching 200°C. The pressure levels and vibration levels are equally as severe. Most manufacturers won't guarantee their components' performance at these levels of severity, so you should analyze each component's technology to determine if there is any part of the design that can't withstand the environment from a technological point of view. For the oil drill operating in a 200°C environment, engineers will need to study the material properties to ensure they aren't getting close to the phase transition point (this is the transformation of a material from one phase of matter to another). You should then test to those levels and beyond to determine how much margin you have. This is basically a component level HALT whereby you must prove that each component can not only survive the environment but also has enough margin in case there are variations in the earth's temperature at those depths, in case your customer decides to sink the drill a bit deeper and expose the drill to even harsher environments.

RELIABILITY INTEGRATION: Integrating Derating with HALT and HASS—COTS Requires HALT and HASS

For many years, the military has been pushing its suppliers to design and build with commercial off-the-shelf (COTS) components. Here is another example of uprating. Many commercial specifications for the components are lower than the requirements for the system. That doesn't mean that the components can't withstand the environment. It just means that the component manufacturers won't guarantee that they do. If you pay them enough money, they will probably build special components (or take standard components and test to a higher standard), but this can be very costly and defeat one of the main purposes of COTS—to reduce the cost of the system. In these cases, you can turn to HALT and HASS as your test vehicles to test in the margin you need. You can use HALT to prove that the design will meet the requirements. You can use HASS to ensure that your manufacturer has a tight enough manufacturing process to ensure that each sample of the component can also meet the requirements.

CASE STUDY: Customized Derating Guidelines

Our client was a power supply manufacturer that needed a derating analysis performed, but the existing derating guidelines weren't suitable for their product. We chose a guideline that best matched their environment, then we tailored the guideline to our client's environment by choosing the components most critical to their application. In this case, the most critical components were the electrolytic capacitors and the power field effect transistors (FETs), which we adjusted the derating guideline for these specific components. We trained the engineers on how to use the guideline. Our client then released the guideline into their documentation system and made part of the design process documentation to use this guideline. Since they started using this guideline, the number of component related failures in the field dropped by more than 25%.

24 Why Do Products Fail?—A Physics of Failure Approach

Physics of Failure (POF) is the process of using knowledge of root-cause failure processes to prevent product failures through product design and manufacturing practices. Using the POF approach, you establish a scientific basis for evaluating new materials, structures, and electronics technologies, which will enable you to uncover information needed to plan tests and screens from which you will determine electrical and thermo-mechanical stress margins.

Using POF techniques, you can model the stress-failure relationship for the dominant environmentally-induced failure mechanisms. Once you develop these relationships, you can then calculate the expected life and compare these results to your requirements. Build POF models to describe a failure mechanism at a particular failure site.

POF models are very different from the traditional Reliability Predictions. Reliability Predictions are sometimes referred to as "black box" models because they model the failures without detailed knowledge of the product. Reliability Predictions use statistical curves to estimate product failures in the field, or they use field data from known sources, such as Telcordia SR332 or MIL-HDBK-217 prediction handbooks.

Dr. Abhijit Dasgupta, professor at the University of Maryland Center for Advanced Life Cycle Engineering (CALCE), states, "Clearly, these prediction models are of value to logisticians who are tasked with supporting existing fielded systems...but they are of questionable value to designers who are tasked with developing new and better technologies." He goes on to draw the following analogy: "Requiring designers to use the empirical MTBF models for building reliable systems is like asking your physician to rely only on statistical mortality tables used by the health-insurance industry when making medical decisions to improve your health!"

RELIABILITY INTEGRATION: Integrating Predictions and POF
Predictions and POF offer two different perspectives on the reliability of your product. Predictions are useful when you want to perform trade-off analyses early in product design, especially when the requirements for making a decision is a relative difference in complexity. POF is better at determining quantitatively what the expected failure rate will be in your particular application. Predictions are sometimes called the "quick and dirty" method, while POF is the more detailed approach. Used together, they make a powerful combination.

CASE STUDY: Using Prediction, End-of-Life (EOL) Analysis, and POF Together on a Micro-Inverter for the Solar Industry
A solar micro-inverter company had a challenging reliability requirement: Their product needed to last 25 years. To meet this challenge, we used three different techniques together—Reliability Prediction, EOL Analysis, and POF.

We started with a Reliability Prediction and determined how close we were to the 25 year requirement based on the Parts Count Analysis method. We then added Thermal Analysis and Derating Analysis and ran a few trade-off analyses to help optimize the design in these two areas.

We then reviewed the components for EOL candidates. These were components that had potential wear-out mechanisms that could come into effect within the 25 year life of the product. We identified three components: electrolytic capacitors, the solder joints, and the glue that held the micro-inverter to the back of the solar panel.

Next, we went through the prediction and identified the components most critical to the design and performed POF modeling on the components to get more accurate failure rate information than we could obtain from prediction databases or vendor data.

We combined all three sets of data into one prediction report and highlighted what areas of the design our client needed to change in order to meet the 25 year requirement.

After our client released the product to the field, we collected field information and compared it to the prediction. We found that it was much more accurate than the previous design in which our client didn't use POF.

25 You Should Design Your Experiments Wisely—The Magic of DOE

Traditional experiments focus on one or two factors at a couple of levels and try to hold everything else constant, which is impossible to do in most processes. When **Design of Experiments (DOE)** is properly constructed, it can focus on a wide range of key input factors or variables and will determine the optimum levels of each of the factors.

If it is unclear which variables contribute the most to the design sensitivity, then it is time to run a DOE to make this determination. In most designs, there are factors that you can control and other factors that are beyond your control. Use a DOE approach to

a. determine which factors are controllable, and

b. help find the optimal balance between each of the factors. This will yield the most robust design possible given the factors present.

25.1 Benefits of a DOE

Here are a few benefits of a DOE:

1. You can evaluate many factors simultaneously, making the DOE approach economical.

2. Sometimes you can't control factors that have an important influence on the output (noise factors), but you can control other input factors to make the output insensitive to noise factors.

3. It isn't necessary that you have in-depth, statistical knowledge to get a big benefit from standard planned experimentation.

4. You can look at a process/design with relatively few experiments. You can distinguish the important factors from the less important ones. You can then concentrate your effort on the important ones.

5. If you overlook important factors in an experiment, the results will tell you they were overlooked.
6. You can run precise statistical analysis using standard computer programs.
7. You can improve quality and reliability without a cost increase (other than the cost associated with the trials).

25.2 Language of DOE

The common language associated with the DOE process:

1. Array—a table in which the factors are across the top (horizontal axis) and the number of runs is on the left side (vertical axis). The size of the array is chosen based on the number of factors and levels.
2. Controllable factors—all of the factors that you have the ability to change.
3. Effects or response—dependent variables (output variables).
4. Factors—independent variables (input variables).
5. Interaction—influence of the variation of one factor on the results obtained by varying another factor.
6. Levels—value at which the factors are set.
7. Main effects—the effects of the factors.
8. Noise—effect of all the uncontrollable factors.
9. Uncontrollable factors—external factors that you don't have the ability to change.

25.3 Steps involved in a DOE

After choosing the DOE you would like to run, here are the steps involved:

1. Select the Factors and Levels
2. Assign Factors to the Array
3. Substitute Values into the Array
4. Calculate the Mean Square Deviation (MSD) and Signal-to-Noise (S/N) Ratio
5. Determine the Correct Combination
6. Build the S/N Response Table
7. Choose the Significant Factors
8. Run Additional Experiments, if Needed

It is easiest to understand the DOE process by using an example, so in Section 25.4, we have run through these eight steps on a simple example of baking cookies.

25.4 Example of DOE – Baking Cookies

Baking cookies from scratch not only depends on the ingredients, it also depends on the cookie size, oven temperature, and baking time. In this example, we will determine the best size, baking temperature, and time for a new recipe. We have a limited amount of time and the judges can only eat a limited amount of cookies. A select panel of judges will rate the resulting cookies on a 0 to 100 scale, where 100 is the best. We average the panelists' results to calculate the final score.

25.4.1 Select the Factors and Levels

After deciding on the DOE you would like to run, the first step is to select the factors and levels. For this, use engineering judgment, history, experience, and previous experiments.

Table 25.1: Selecting Factors and Levels

Factor	Level 1	Level 2
A: Oven Temperature (in °C)	325	375
B: Cooking time	12 minutes	15 minutes
C: Cookie size	Small	Large

25.4.2 Assign Factors to the Array

The next step is to choose the size of the array based on the number of factors and levels. Table 25.2 shows the L4 array we used for this cookie DOE. An L4 array indicates that we will need four runs for the experiment. It is the appropriate array size for a three factor/two level experiment such as this. Next, we build the array and assign levels to each cell, as shown in Table 25.2.

Table 25.2: L4 Array for Cookie DOE

L4			
Run No.	**A**	**B**	**C**
1	1	1	1
2	1	2	2
3	2	1	2
4	2	2	1

25.4.3 Substitute Values into the Array

Next, we take the values from the experiment and substitute them into the array. Table 25.3 is the L4 array with the experimental values inserted.

Table 25.3: L4 Array with Experimental Values Inserted

L4 (2^3)			
Run No.	**Temp**	**Time**	**Size**
1	325	12	Small
2	325	15	Large
3	375	12	Large
4	375	15	Small

25.4.4 Calculate the MSD and S/N Ratio

The next step is to calculate the MSD and the S/N values using the following formulas:

$$MSD = \frac{\frac{1}{Y_1^2} + \frac{1}{Y_2^2} + \cdots + \frac{1}{Y_n^2}}{n} \qquad \textit{Formula 25.1}$$

$$S/N = -10\log(MSD) \qquad \textit{Formula 25.2}$$

Table 25.4:   Calculating MSD and S/N

Run	A	B	C	Y1	Y2	ΣY	Avg Y	MSD	S/N
1	1	1	1	69	62	131	65.5	0.000235	36.29
2	1	2	2	38	37	75	37.5	0.000711	31.48
3	2	1	2	39	41	80	40.0	0.000626	32.03
4	2	2	1	26	23	49	24.5	0.001685	27.73

25.4.5 Determine the Correct Combination

We could select the best of the four combinations, yet that would be ignoring the ability to make a selection from all possible combinations. Instead, with a little math we can determine the right mix of time, temperature, and size for the highest scoring cookies.

Table 25.5:   Determine the Correct Combination

Factor	Level	ΣY	Y	S/N
A	A1	131 + 75	51.50	33.88
	A2	80 + 49	32.25	29.88
	Total		83.75	
B	B1	131 + 80	52.75	34.16
	B2	37.5 + 24.5	31.00	29.61
	Total		83.75	
C	C1	131 + 49	45.00	32.01
	C2	75 + 80	38.75	31.76
	Total		83.75	

25.4.6 Build the S/N Response Table

Next, we reduce Table 25.5 to build the S/N Response Table to determine which factors are most significant.

Table 25.6: S/N Response Table

Level	Factor A	Factor B	Factor C
Level 1	33.88	34.16	32.01
Level 2	29.88	29.61	31.76
Difference	4.00	4.55	0.25

25.4.7 Choose the Significant Factors

Now that we have calculated the differences in S/N, we can choose the significant factors. As we can see from Table 25.7, the significant factors are those with large S/N differences. Based on this, we conclude that oven temperature is a significant factor, and the lower level is better than the higher level. We also conclude that cooking time is a significant factor, and the lower level is better than the higher level. Finally, we conclude that cookie size *isn't* a significant factor, and therefore we don't need to worry about controlling the size of the cookies.

Table 25.7: Choose the Significant Factors

Factor	Level 1	Level 2	Reason
A: Oven Temperature (in °C)	325	375	Significant difference (>3dB). Select larger S/N
B: Cooking time	12 minutes	15 minutes	Significant difference (>3dB). Select larger S/N
C: Cookie size	Small	Large	Slightly higher S/N, could go either way

25.4.8 Run Additional Experiments, if Needed

Now that we know the two factors that are most significant and we know the approximate amount of temperature and time, we can run additional experiments to determine more accurately the optimal temperature and time.

RELIABILITY INTEGRATION: Integrating DOE with HASS—DOE Can Be Used When Identifying How to Combine Stresses for HASS

It may be worth running a DOE with different stresses to find which stresses combined together are best at finding defects for HASS. With HASS, you have many different potentially effective stresses available. Some of these stresses are fast temperature cycling, slow temperature cycling, steady temperature, vibration, humidity, voltage margining, and power cycling. How do you know which stresses are best for your manufacturing environment, and how do you know which stresses to run by themselves versus combining with other stresses? You need to make a decision with the fewest production samples as possible. Therefore, DOE is a great technique for this. You can use the different stresses as the factors and conduct the experiment by subjecting a set of samples to each of the stresses. Depending on the failure rate during manufacturing, you will need to choose a sample size high enough to ensure that your sample population has a defect. Alternatively, you can purposely inject known defects into your sample population (we call this "seeding," and we describe this process more in Section 43.2.6.2.2). You can then measure which stresses were most effective by the number of defects found.

CASE STUDY: DOE on a Sleep Apnea Product

A company making a sleep apnea product came to us with an issue on their new product in which the blower assembly was getting noisy after about six months in the field. Our client used HALT during development but didn't uncover this issue. In HALT, the blower assembly performed well beyond its rated specification without showing any sign of noise.

Using Design of Experiments (DOE), we identified the eight variables that most affected performance of the blower assembly. We then had the blower manufacturer build different versions of the blower with these different variables in place. Next, we subjected the blower to an accelerated temperature test and monitored acoustic noise output and motor vibration.

After only thirty days of accelerated testing, we were able to identify optimal design parameters for the blower assembly. Our client made our recommended design changes and released the product for volume manufacturing. It has been running for over three years nearly free of defects.

Chapter

26 Don't Tolerate Tolerance Issues

The nominal value specified on an engineering drawing also includes a **tolerance** with a high side and low side value. Those high and low side values indicate how much variability to allow for that component part (or function). These values can be for both functional characteristics, such as force, velocity, or temperature, and for geometric characteristics, such as length, parallelism, or circularity. There are two decisions that you need to make once you decide on the characteristics to specify:

- What is the best nominal value of the characteristic to assure robust low cost performance?

- What variability levels to allow in the design to balance cost and quality?

Engineers have many methods to help decide nominal values, and equally many ways to specify tolerances. Some are good and some aren't so good. The consequences of selecting improper nominal values and tolerances usually include costly countermeasures downstream. Specification of tolerances that are too tight may require very precise costly manufacturing processes or secondary operations. Specification of tolerances that are too loose may incur quality problems during assembly or functional problems downstream, either in manufacturing or field service. Reliability issues also arise from deterioration of design settings over time from wear and tear.

Amusing Anecdote: I was teaching a reliability class to a company that manufactured hot tubs. When I got to the section on **Tolerance Design**, I mentioned to the class that I actually owned one of my client's hot tubs, and I noticed that the flap that slides up and down to allow water to flow to the filter compartment started sticking over time. I told the class that this could be an issue with improper tolerances because something in the design drifted over time and caused it to start sticking. Just then, one of the engineers in the back of the room stood up and said, "I'm so tired of hearing about this issue—I fixed it already." After the class, another of the engineers told me that this was one of the highest failure items in the field.

Noise factors, like manufacturing variation, materials variation, or other sources, often affect important functional variation of a design. You often need to control noise factors to within narrower ranges, even though the design cost may rise. For noise factors that aren't so important to functional variation, you should widen the tolerance range, thus potentially reducing costs. Both tolerance widening and tightening tolerances are addressed with Tolerance Design. Tolerance Design was originated by Dr. Taguchi in Japan fifty years ago and was used to apply experimental design techniques to decisions (to improve a design) where cost and quality are balanced.

Tolerance Design uses Design of Experiments (DOE) techniques, along with economic considerations, to control the output variation of a design. In Tolerance Design, assign noise factors and their levels to an orthogonal array. The orthogonal array enables fair comparison of tolerance factor main effects. Advanced statistical tools, like Analysis of Variance (ANOVA), enable estimation of fractional contributions and error variance. Assign two or three level factors to an orthogonal array, following standard layout practice. In a two-level design, such as L12, you can assign up to eleven two-level tolerance factors. The two levels you assign are ideally the +/- 1 limits for the factor (but this isn't a requirement). Assign three level factors to an orthogonal array, such as an L18, in the same manner with the middle level at the nominal value.

These levels give similar answers to those you would obtain using normal distribution Monte Carlo simulation. Set the levels of the noise factors according to the orthogonal array fairly accurately. In an L12 assignment, there are twelve distinct combinations of factor tolerances in the layout. In an L18, there are eighteen distinct combinations of factor tolerances in the layout. A systematic experimental design like L12 or L18 allows you to unravel the effects of each tolerance on the output variation.

If actual hardware isn't available for Tolerance Design experimentation, but a reasonable physical model is readily available, then assign tolerance levels to the factors in the model based on your current understanding of the design.

RELIABILITY INTEGRATION: Integrating Tolerance Design with DOE
Noise factors, like manufacturing variation, materials variation, or others, often affect both design cost and important functional variation of a design. Use DOE with Tolerance Design to control the output variation of your design, using economic considerations.

Chapter 26: Don't Tolerate Tolerance Issues

CASE STUDY: Printer from a Fusing Subsystem Model

This printer example comes from a fusing subsystem model where we assigned 27 three-level tolerance factors to an L81 orthogonal array. The fusing system was used to transport paper and toner through cylindrical elastomeric heated rollers, to melt the toner into the paper, and to release the sticky toner from the rollers when fusing was complete. Factors included roller velocities, mechanical and thermal material properties, shape parameters, and a few others. The response of interest was the dwell time in the heated zone between the two rollers. If the dwell time was too long or too short, quality problems occurred. Some tolerances in the system had very large effects on the dwell time, while other tolerances in the system had very small effects on the dwell time. The Tolerance Design modeling experiment constructed 81 different combinations of system tolerances and estimated the dwell time for each. The results indicated that we needed to narrow the tolerances for several factors; there were also opportunities for us to widen other tolerances to reduce cost. The overall improvement was an approximate thirty percent reduction in standard deviation of dwell time with no added cost.

27 Preventing Failures *before* They Happen

Wouldn't it be great if you could predict all failures *before* they happened, then change the components just before the failures occurred? This is called Prognostics, and in some applications, we as an industry aren't far from being able to do this. In 2008, the Institute for Electrical and Electronic Engineers (IEEE) Reliability Society started holding an annual conference on Prognostics and Health Management (PHM) to address this very issue. Right now, Prognostics is being applied primarily on high-end expensive systems requiring high reliability and availability, but this may soon change with the help of conferences such as PHM.

In some industries like the telecom industry (where the allowable downtime is measured in seconds per year) and the oil exploration industry (where each failure down-hole can cost more in lost production than equipment costs), companies are willing to pay for this innovation to prevent impending failures.

Engineers have been applying prognostics to mechanical components for many years. For example, on motors, we can monitor for changes in noise and current. However, most industries building systems with mechanical components choose not to use prognostics because of the cost of the instrumentation within the product needed for this. With advances in technology, the cost of the instrumentation is dropping. Recently, we have started to discover clever ways to do this with electronics. Measuring fluctuations in power draw, temperature, clock frequency drift, and other parameters can indicate an impending failure. It takes experience to know what parameters to monitor and how much of a change in each parameter would be considered a failure. When you can't determine when a particular failure is impending, you must use past test data to dictate when to replace wear-out items. This is called Preventive Maintenance.

27.1 Preventive Maintenance

Preventive Maintenance (PM) assumes that your product has components that wear and must be replaced on a periodic basis. Preventive Maintenance can be based on:

- Scheduled service for cleaning
- Service for lubricating
- Detection of early signals of problems
- Replacement after specific length of use

It is important to determine how often to replace a component. If you don't replace often enough, you risk the possibility that the component will wear out and cause a system failure. Fans are a good example of a wear-out failure that can potentially cause a system failure. If you replace a component too soon, then you increase the opportunity for failure during the maintenance. Whenever you replace a component, there is a possibility that the new component fails prematurely or that you introduce a failure during the maintenance action (e.g., cleaning and lubricating). By replacing a component too soon, you are increasing the number of opportunities for these two possibilities to occur.

Let's take your oil filter on your car to illustrate PM. Your oil filter needs to be replaced on a periodic basis, and your owner's manual will instruct you on the interval. If you wait too long, you risk damaging your engine. If you replace it ahead of schedule, there is a possibility that the person performing this change will install the incorrect oil filter or will install the oil filter improperly. The more often you change your oil filter, the higher the likelihood of this occurring. In addition, by replacing it more often, you are incurring more expenses. The best scenario is to follow your owner's manual and replace it within the prescribed interval—no more and no less.

27.2 Maintainability

In other industries, only rudimentary Preventive Maintenance techniques are deployed, such as replacing wear items on a regular basis (e.g., fans and belts), and for the rest of the failures, the focus is more on speed of maintenance. This is called Maintainability Analysis. Maintainability Analysis is the method of determining how to best design a product for ease of maintenance. A Maintainability Prediction is the method of determining how long it takes to repair a product once it fails. A Human Factors Analysis is the study of all aspects of the way humans relate to the equipment, with the aim of improving operational performance, reliability, and safety. A Maintainability Analysis is closely related to a Human Factors Analysis because a large part of the repair process involves the user. During this analysis, analyze your product for ease of maintenance and calculate the mean time to repair (MTTR) for your product. The MTTR is the measure of the average repair time. This is calculated by averaging the repair times for each individual replaceable assembly, weighted by the failure rate. The main components that go into this calculation:

1. Time to detect and isolate a failed assembly
2. Time to get to the failure location
3. Time to replace the assembly
4. Time to validate that the repair was successful

27.3 Availability

Availability differs from Reliability in that with Availability, you are concerned with the system being up and operational. In Reliability, every failure counts. In Availability, the failures only count if the user experiences them. Availability is the probability a system is ready for use when needed or the proportion of time a system is ready for use. The formula for Availability is:

Availability = MTBF / (MTBF + MTTR)   Formula 27.1

Where:
MTBF is the mean time between failures
MTTR is the mean time to repair

The case study at the end of this chapter highlights the difference.

27.4 Reliability, Availability, Maintainability, and Preventive Maintenance Trade-Off

What happens when you need to be concerned with Reliability, Availability, Maintainability, and Preventive Maintenance in the same product? Are there any industries that focus on all four simultaneously? How about the auto industry? If your car isn't reliable, you will pay a lot of money in repairs. If it isn't highly available, you will be late frequently for appointments. If it isn't easily maintainable, you will spend a lot of money every time it needs to be serviced. If there are no good preventive maintenance guidelines, then you will wait until components wear out rather than replacing components on a routine basis. As a result, your car may suffer more damage.

The problem is that each of these disciplines may have conflicting strategies. What do you do then? You need to optimize. Open up the hood of your car and look at where the manufacturer placed the different engine components. Why did they place them where they did? Well, partly for functionality of course, but also to optimize maintenance. A good automobile manufacturer will estimate which items are likely to fail more often and make those more accessible. They position the more reliable components so that they are less accessible. Your preference likely would be to make everything more accessible, but this would increase the size of the car and lower its fuel efficiency, so this isn't a realistic option.

This is no different from your computer. Some of you have been in the industry long enough to remember the first generation of personal computers. The hard drive was always easy to access as was the power supply. However, to replace the motherboard, you had to take out every other component in the chassis. Why would the chassis manufacturer do that? Well, they figured that the hard drive and power supply would fail more often, and you know what? They were right (and twenty-five years later, the same is true).

RELIABILITY INTEGRATION: Integrating Preventive Maintenance with Testing—Determining a Preventive Maintenance Interval Using Test Data
Nothing will drive the cost of ownership of a product up faster than if the manufacturer suggests a higher than necessary (or lower than required) interval of Preventive Maintenance. Optimization is the key. See the example in Section 27.1 of this chapter on why you should determine the Preventive Maintenance interval accurately. The only way to arrive at this optimization is through testing, which is one of the uses for an Accelerated Life Test (ALT).

CASE STUDY: Confusing Reliability with Availability

Our client was a manufacturer of inverters for the wind energy market. They hired us to perform a Reliability Prediction to meet their customer's requirement of an MTBF of ten years. When we questioned the rationale behind the number, our client's customer told us they needed the product operational for ten years. We explained to them that an MTBF of ten years doesn't mean the product will operate for ten years. We told them that if the failures are random, then more than sixty percent of the product will fail within 10 years. Using Formula 27.1, we can calculate that 63.2% of a population will fail within the MTBF time interval (if =1/t, then e^{-1}=63.2%).

$$R=e^{-\lambda t} \hspace{4cm} \textit{Formula 27.2}$$

where:
λ = 1/MTBF = failure rate
t = product life

Our client's customer was clearly confused. After a long silence, they explained to us how they arrived at the figure of ten years. The told us that in the wind industry, the wind only blows about thirty percent of the time (or only blows strong enough thirty percent of the time to allow the blades of the turbine to spin). During the time the wind is blowing, our client's customer wanted the product working one hundred percent of the time (or as close to that as possible) for ten years. In the other seventy percent of the time, they would allow the maintenance personnel to check out the system and perform routine Preventive Maintenance on the system. From this, we realized our client's customer was specifying Availability, not Reliability. We explained this to our client and his customer and convinced them to change their goal to an Availability goal.

FIRST RULE OF A GOAL STATEMENT: You have to be careful what your customer asks for in terms of goals because they may not know what they really want unless you ask the right questions.

At this point, we restated the goal in terms of Availability and then embarked on a redesign to add in Prognostics into the circuitry to detect when specific components would start to wear. With this new design, we were able to meet the Availability goal for the product.

28 What Happens When Your Product Reaches Its End of Life?

End-of-Life (EOL) Analysis estimates the onset of wear-out failures for consumables (e.g., electrolytic capacitors, fans, motors, and drives). This can aid in planning Warranty Costs and scheduling Preventive Maintenance (PM).

28.1 End-of-Life (EOL) Prediction

During the design phase, review your Reliability Prediction and FMEA to determine which components in your system have a dominant wear-out mechanism. For these components, perform an EOL Prediction. Use data from your Accelerated Life Tests to help in this analysis. See the Reliability "Bathtub" Curve in Figure 28.1 for where the onset of end-of-life occurs.

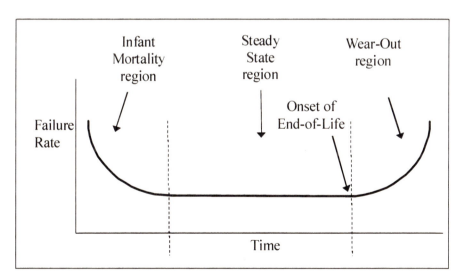

Figure 28.1: Reliability "Bathtub" Curve

28.2 End-of-Life (EOL) Analysis

After you start shipping your product, plot field data against time to failure data to determine if the product is approaching the wear-out region. Then, compare this with the predicted estimate for the EOL to determine confidence in results. Next, make an estimate as to when the end-of-life will be, along with the slope of the end-of-life curve. See the Weibull Plot in Figure 28.2 for an example of a product entering its end-of-life phase. This is indicative of a Beta (β) > 1. In this example, β= 2.9. This parameter β describes how fast the population will wear out once wear-out begins on the first sample. If wear-out is present, then β will be greater than one but how much greater? The greater the value of β the faster the wear-out occurs. A β of 1.5 indicates a gradual wear-out, whereas a β of 4 indicates a very sudden wear-out. From this, examine the cost impacts and make recommendations for when to stop repairing failed products and when to start offering upgrades and trade-ins. Also, compare specific failure trends with predicted trends and perform corrective actions for those trend items.

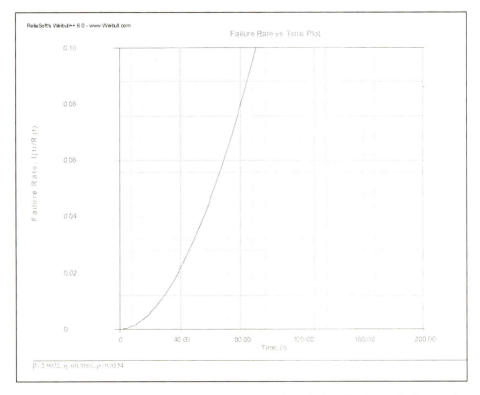

Figure 28.2: Weibull Plot (using ReliaSoft's Weibull++ Software)

RELIABILITY INTEGRATION: Integrating an EOL Analysis into Design Changes

After performing an EOL Analysis, you may find that you need to change a component in your design, especially if the component isn't fit for the end-use application. We find this with fans and hard drives all the time, especially when designers don't match the rated duty cycle of the component to the expected duty cycle in the application. When this occurs, you get wear-out failures. Your EOL Analysis will point out these mismatches between rating and application.

CASE STUDY: Modifying Preventive Maintenance Strategy Based on End-of-Life Results

A semiconductor manufacturing equipment company had instituted a five-year Preventive Maintenance (PM) schedule on their belts. They asked us to review their field data to find out how many failures they were getting in the five years to determine if they needed to change their PM time. We reviewed their PM data as well as the life test data and determined that the five years initially set forth was too optimistic. The data indicated that the components were wearing too much in between PM cycles, and failures were resulting. We recommended that they lower their PM interval to every three years. Our client was able to make this change early in the life of the product before a significant population of product was in the field.

29 Think About Warranty Now

Warranty Analysis is part of the Design for Warranty (DFW) methodology, in which you use the goals, strategies, and data to jump-start your analysis with the development team during product design. The objective is to identify and prioritize the appropriate warranty metrics, goals, strategies, and action plans to reduce warranty expenses. Optionally, you can perform an in-depth warranty cost analysis for a released product to better estimate potential warranty cost savings.

DFW requirements:

1. Identify the most important (costly) warranty events you need to address.
2. Make explicit choices of support process used to resolve a warranty event.
3. Assign estimated warranty dollars to FMEAs.
4. Identify needed supportability requirements.
5. Identify product serviceability requirements.
6. Identify product maintainability requirements.
7. Improve supply chain processes.

DFW is uniquely different from Design for Reliability (DFR), Design for Supportability, and Design for Maintainability. DFW isn't just another deliverable to add to the product life cycle's checklist. One common technique between all of these is FMEA. FMEA is the best analysis technique when performing Warranty Analysis because the more accurately you can predict which events will occur and how often they will occur, the more accurately you will be able to estimate warranty costs. See Figure 29.1 for a view of the interaction between DFW and DFR.

Figure 29.1: Relationship of Design for Warranty (DFW) to Design for Reliability (DFR)

DFW is a set of models, frameworks, and techniques you can use as an integral part of the DFR process to reduce warranty costs.

29.1 Definition of Warranty

Warranty is a guarantee, given to the purchaser by a company, stating that a product is reliable and free from known defects and that the seller will, without charge, repair or replace defective parts within a given time limit and under certain conditions.

29.2 Types of Warranties

The two types of warranties are product warranties and extended warranties.

Attributes for product warranties:

1. Product warranties start at the time of purchase.
2. Product warranties are either implied or explicit.
3. Financial liability for product warranties is with the original equipment manufacturer (OEM).
4. The length of product warranties is country-dependent (e.g., for the European Union, product warranty is greater than one year for consumer products).
5. Product warranties are expensed and paid for by the warranty reserve. The length and coverage are subject to country regulations. The accruals as well as the amount spent on warranty is required to be reported in the USA.

Attributes for extended warranties:

1. Extended warranties are treated as trade orders for contracted support "services" (customer pays extra for it).
2. Extended warranties may be sold and administered by third parties.
3. Expenditures for extended warranties don't have to be broken out from ordinary operating expenses.

Both product warranties and extended warranties affect your customers' perception of quality, but they do so differently. Consumers usually expect a product to last as long as the product warranty and will view a product as having poor reliability if it fails within this period. However, consumers don't always expect a product to last into its extended warranty. The consumers' perception often depends on whether they purchased an extended warranty contract. If so, then often there is no ill will towards the manufacturer. However, if the consumer did not purchase an extended warranty, there could be.

29.3 How Big Are Warranty Costs?

In the USA, the warranty costs for all publicly traded companies was greater than $28 billion in 2008. The computer and high technology sector accounted for $8 billion of this. This varies from 0.5% to > 8.0% of revenue (excluding "service" revenue). This is close to the percentage of revenue spent on research and development in some low-margin businesses. Industry-wide, warranty costs as a percentage of revenue is increasing year after year.

29.4 The Warranty Burden

Warranty burden isn't uniformly distributed throughout the design and manu-facturing supply chain. In the supply chain, you have contract manufacturers (CMs), original design manufacturers (ODMs), and component suppliers. OEMs are starting to ask their supply chain to share the burden of warranty costs.

29.5 Warranty as a Metric

The warranty metric is becoming the publicly available metric for a company's product quality in several different ways:

1. Warranty is being used by industry press as a quality metric.
2. Warranty is being used by management teams as the business metric for quality.
3. Warranty is being used by companies as a quality benchmark with their competition.

29.6 Warranty Event Cost Model and its Application

Types of warranty support processes:

1. Home repair
2. Onsite repair
3. Bench repair
4. At an authorized service provider
5. Centralized (mail-in)
6. Self-repair (customer replaces a part)
7. Phone support

See Formula 29.1 for the warranty cost model formula. You can use this formula to calculate the total expected warranty costs. The processes that you need to include in the model are those processes used to resolve warranty events and warranty-related issues. Include the following:

1. Call center processes (even if you outsource this function)
2. Support organization processes (even if you outsource this function)
   a. Onsite repair by service engineers or authorized service suppliers (ASPs)
3. Supply chain processes
   a. Spare parts management (usually as an overhead)
   b. Spare parts logistics
4. Manufacturing/factory resources expensed against the above-referenced processes (usually as indirect/overhead expenses)
   a. Return parts testing process
   b. Reliability engineering resources

$$Cost_{Total} = \sum_{i=1}^{N} f_i * C_{STDi}$$
Formula 29.1

where:

$Cost_{Total}$	=	the total cost of warranty for this product
f_i	=	frequency of occurrence of the ith warranty event type
N	=	total number of warranty event types
C_{STDi}	=	standard service process cost utilized to resolve the ith warranty event type (including direct material costs)

You can use this model in two ways:

1. Use actual warranty cost data to develop a model for a similar or current product. For this, you can either:
   a. Use actual data from a current or similar product (sampling schemes are often used for large data sets)
   b. Use actual data for the top 10–15 Pareto Warranty events and then combine it with new product's high probability warranty events (failure modes from FMEA).
2. Use existing product information to establish standard costs for planned support processes. Build cost model based on FMEA to identify high probability warranty events.

For each warranty event, develop design alternatives using estimated warranty cost savings as one of the evaluation measures.

29.6.1 Making Warranty Cost Reduction Choices

Using the Warranty Event Cost Model, you can reduce warranty costs by reducing either the frequency of occurrence or the cost per occurrence.

1. Methods to reduce the frequency of occurrence
   a. Increase hardware reliability
   b. Improve software/firmware robustness
   c. Improve fault tolerance for specific failure modes
   d. Change/remove product features
   e. Change how product works
2. Methods to reduce the event's process cost:
   a. Design new (cheaper) processes
      * Design process around new technology
      * Design process to meet new market needs
   b. Switch to a cheaper process
      * ID features/capabilities needed to support different process
   c. Reduce standard process costs by:
      * Product changes
      * Process improvements
      * Outsourcing
      * Supply chain re-engineering

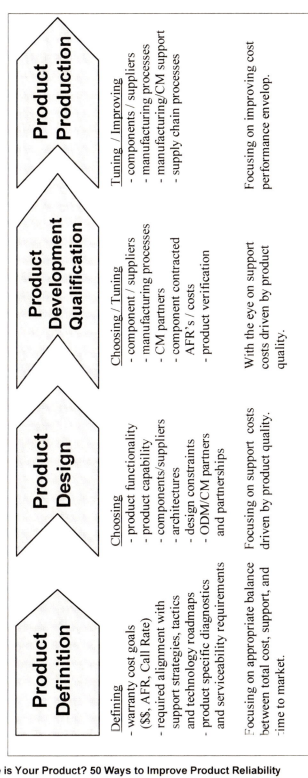

Product Definition

Defining
- warranty cost goals ($$, AFR, Call Rate)
- required alignment with support strategies, tactics and technology roadmaps
- product specific diagnostics and serviceability requirements

Focusing on appropriate balance between total cost, support, and time to market.

Product Design

Choosing
- product functionality
- product capability
- components/suppliers
- architectures
- design constraints
- ODM/CM partners and partnerships

Focusing on support costs driven by product quality.

Product Development Qualification

Choosing / Tuning
- component / suppliers
- manufacturing processes
- CM partners
- component contracted AFR's / costs
- product verification

With the eye on support costs driven by product quality.

Product Production

Tuning / Improving
- components / suppliers
- manufacturing processes
- manufacturing/CM support
- supply chain processes

Focusing on improving cost performance envelop.

Figure 29.2: Decisions that Drive Warranty Costs across the Product Life Cycle (PLC)

29.6.2 Warranty Event Cost Model Case Study

We have included a case study to highlight the use of the Warranty Event Cost Model.

Business:	Large commercial computer product developer. High volume, moderately priced products.
Situation:	Warranty bill for business > $10M/yr range; warranty @ ~ 4.1% of revenue. Support services performed by ASPs. Major support processes ranged from on-site repair to phone support. Only evolutionary products, no revolutionary products.
Applying the Cost Model	Top 15 warranty issues accounted for > 80% of warranty costs. Top 15 came from current product performance analysis *and* new product FMEA. Process analysis confirmed 4 primary support processes used for > 95% of warranty events. Could make reasonable estimates at standard process costs regardless of poor information technology (IT) infrastructure.
Using model to make design trade-offs	Used model to make design trade-off decisions on 8 of 15 top warranty event cost contributors. Total planned warranty cost saving: from 5.5% to 2.6% of expected product revenue.
Results	All needed additional product features, functionality and diagnostic tools implemented and deployed at NPI. One year after NPI, company exceeded their planned cost savings

29.6.3 Cost Reduction Strategies across the PLC

Primary warranty cost drivers that product development can directly affect:

1. System Reliability (AFR)
2. Delivery Mix
3. Service Process Costs used by significant warranty events
4. Supportability features and functions of the product (e.g., diagnostics, embedded tools)
5. Serviceability of product
6. BOM choice effects on supply chain (both manufacturing and spare parts)
7. Supplier Cost Recovery

RELIABILITY INTEGRATION: Integrating Warranty Events with FMEA
FMEA is the best analytical technique when performing Warranty Analysis because the more accurately you can predict which events will occur and how often they will occur, the more accurately you will be able to estimate warranty costs. FMEA typically uses the three attributes severity (S), probability (P), and detection (D) as the measure of risk for each failure mode. When designing for warranty, we recommend switching from S, P, and D, to S, P, and C, where S is the severity of the failure mode, P is the probability of the failure mode occurring, D is the likelihood of detecting the failure mode, and C is the cost of the failure mode if it does occur.

CASE STUDY: Warranty Analysis in the Fast Food Industry
A fast food company wanted to change a portion of their restaurants from typical sixteen-hour-a-day operations to 24/7 operations (in which they would never close the restaurant). This meant that their equipment would be utilized for longer periods of time each day with virtually no off time. We reviewed their point-of-sale equipment to determine if any component(s) in the product would be adversely affected by this change. Simple reliability analysis showed that going from sixteen hours a day to twenty-four hours a day would decrease the life of the electronics by fifty percent. However, one component in particular, the hard drive, would degrade much faster because typical hard drives for laptop and desktop applications require a periodic cool-down cycle period. In the case of laptops, this usually comes at night when the user is asleep—the user may leave the laptop on, but the drive is usually not being used. We consulted the drive manufacturer, and they told us that in this new application, the five-year warranty wouldn't be valid. They said that the company could expect closer to one to two years of operation in this new environment. They recommended that our client move to an industrial-grade hard drive built for 24/7 operation. Our client implemented this recommendation and the new drive met its expected field reliability.

30 Infinite Possibilities with FEA

Finite Element Analysis (FEA) is a technique used by engineers to estimate the responses of structures and materials to environmental factors, such as fluid flow, forces, heat, and vibration. It is possible to model complex mechanical components by subdividing a component into small, "finite" elements, and analyzing the component as an assembly of these small, simple elements. You can use Finite Element models for a wide variety of applications, such as determining the natural frequencies and mode shapes of a structure, estimating the dynamic response, or determining the stress distribution due to a generated thermal load.

An FEA consists of five main steps:

1. Begin the process with the creation of a geometric model. Use an existing computer-aided design (CAD) model where it is available.
2. Break the model down into smaller elements of simple shapes connected at specific node points. This process is called "meshing." The result is a finite element mesh of each component.
3. The material properties are prescribed for each component. These can be simple elastic properties for routine predictions, or these can utilize more complex material models, such as elastic-plastic, elastomeric, and creep.
4. Apply any boundary conditions (loads) to the model. The boundary conditions may include a combination of specified displacements, forces, pressures, and thermal loads. Contact conditions are often important to allow for more general interaction, relative sliding, and separation between bodies.
5. Finally, perform the FEA by solving the system of equations that result from the particular model and boundary conditions. Display the results graphically or through tabular reports to aid in understanding the output and identifying the critical regions.

Commercial software packages, such as MSC.Nastran or MSC.Marc[4], ANSYS[5], Abaqus FEA[6], are available and offer a broad range of analysis capabilities and options. You can perform linear analysis for initial design, but linear analysis is limited to small deformations and elastic material response. Nonlinear analysis is quite general and includes the full spectrum of effects, such as large displacement, contact, large strain, and nonlinear material response. While linear analysis is valuable, it is important to recognize its inherent limitations. The common phrase "nature is nonlinear" shows the importance of the more general, nonlinear regime for many applications.

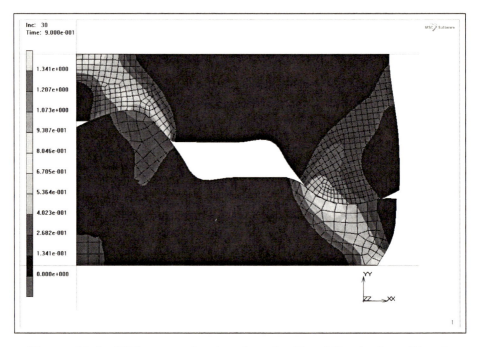

Figure 30.1: FEA example showing the Total Equivalent Plastic Strain for a component

---

4. MSC.Nastran and MSC.Marc are from MSC Software Corporation
5. ANSYS is from ANSYS, Inc.
6. Abaqus FEA is from Dassault Systems

RELIABILITY INTEGRATION: Integrating FEA with ALT—FEA Can Help Plan an ALT

When you are planning an Accelerated Life Test (ALT) for a mechanical assembly, you should understand the failure mechanism(s) you are trying to accelerate. Otherwise, you may choose inappropriate stresses or stress levels. An FEA is great for understanding these failure mechanisms and their relative importance. Once you complete the ALT, you should then feed the information back into your FEA model to compare for accuracy of the model and make adjustments as needed so that the model is even more accurate on future ALTs.

CASE STUDY: FEA Used to Model a Drop Test

We performed a non-linear FEA to simulate a drop test on a mobile phone. The purpose of the analysis was to determine how the product would fail during the design process so that our client could make changes to the product before developing molds and manufacturing tools. Once they develop molds and tools, it is much more expensive to make mechanical changes to the product. First, we met with the client to determine what components were of most interest. We then reviewed our client's mechanical engineering drawing package. Next, we performed the meshing effort. We defined material properties, applied appropriate boundary conditions, and built connections between components in our FEA software tool. We then executed the FEA model by subjecting the model to the various drop heights and directions.

Next, we performed lab stiffness testing to supplement the material property information we were able to gather from industry sources. We then created a lumped mass model (LMM), lumping the assembly into a spring-mass-damper representation. This simpler LMM is useful for making design decision and eases interpretation of complex FEA results.

Finally, we performed trade-off analyses with the client, making changes and seeing how it affected the overall performance. Using this approach, our client was able to make several design improvements and optimize the design before building the first sample.

31 Software Reliability (No, This Isn't an Oxymoron)

With all the complaints you hear about products rebooting and software crashing, do companies really practice Software Reliability? In fact, there are some companies that do, but they are mostly in the industries that require products to have high availability, such as telecom, defense, and space, or safety-averse industries, such as medical and industrial plant operation. Most other industries don't pay as much attention to it.

In this chapter, I will show you easy ways you can practice Software Reliability even if you aren't in one of these industries.

31.1 Software Reliability Background

I estimate that **Software Reliability** is about twenty years behind Hardware Reliability as measured by the percentage of companies that have embraced the methodology. Much of the reason for this is due to the following:

1. Ramifications of failure—many software bugs only require a reboot, and most users complain less about reboots than hard failures. Our experience has shown that software failures occur ten times more frequently than hardware failures (counting critical bugs only, as well as bugs found by the testers before shipping a release), but they largely go unreported.

2. Education on the consumer side—many consumers have just come to expect the fact they need to reboot their product periodically, typically due to memory leaks. Memory leaks occur because software routines use up a portion of memory when they load a program, then don't give the memory back after they close the program, which causes the available memory to shrink. This in turn results in the product slowing down, and the only way the consumer can overcome this is by rebooting the product.

3. Education on the manufacturer's side—education here is two-fold. Many manufacturers don't know some of the new innovative methods for improving Software Reliability (many of these methods are described further in this chapter), and the manufacturers often don't take the time to determine how the user may use/abuse their product.

4. Software engineers tend to be a bit free-spirited. That's not a bad thing—in fact, that is what drives much of the innovation. However, along with this comes a sense of rebellion against structure and processes.

5. The entry cost for a software development team is much less than for a hardware team, so software companies are springing up in countries all around the world—Russia, India, China, and Vietnam, to name a few. This is driving down the cost, and when companies compete primarily on cost, reliability and quality tend to suffer.

31.2 Software Reliability Approaches

For most companies developing products containing embedded software, their Software Reliability programs aren't effective enough to produce products that meet customers' expectations. We can group these programs into three different categories:

1. Software Predictions
2. Software Process Control
3. Quality through Software Testing

31.2.1 Software Predictions

Traditional Software Reliability programs treat the development process as a software-generating black box. Teams develop software prediction models to provide estimates of the number of faults in the resulting software; greater consistency in reliability leads to increased accuracy in the modeling output. Within the black box, teams use a combination of reliability techniques, such as failure analysis, defect tracking, and operational profile testing to identify defects and produce Software Reliability metrics. However, the metrics required for accurate predictions are difficult and time-consuming to collect accurately, and if this data isn't accurate, it limits the usefulness of the prediction. In addition, predictions can only measure where you are; they don't have the ability to show you where to make improvements.

31.2.2 Software Process Control

Software process control assumes a correlation between software process maturity and latent defect density in the final software. If the current process level doesn't yield the desired Software Reliability, companies implement audits and stricter process controls. Process control is a good first step, but it doesn't in itself guarantee reliable software (this is analogous to ISO9000 programs—documentation is necessary but is only the first step).

31.2.3 Quality through Software Testing

Quality through software testing is the most prevalent approach for implementing Software Reliability within small or unstructured development organizations. This approach assumes that organizations can increase reliability by expanding the types of system tests (e.g., integration, performance, and loading) and increasing the duration of testing. Organizations then measure Software Reliability by various methods of defect counting and classification. Generally, these approaches fail to achieve their Software Reliability targets. Companies find that their software engineers spend more time debugging than designing or coding, and accurate Software Reliability measurements aren't available at deployment to share with customers.

31.3 New Software Reliability Approach— Design for Software Reliability

The best method to increase Software Reliability without significant increases to schedules or budgets is to use a Software Design for Reliability (SDFR) approach, similar to methods we have been discussing for hardware. Sections 31.3.1 through 31.3.5 describe the Software DFR Process.

31.3.1 Software Reliability Assessment

Before starting a Software Reliability program, perform a Software Reliability Assessment by assessing your team's capability to produce good software. Use the same assessment methods described in Chapter 5.

Benchmark your development practices against industry best practices to ensure they have a solid foundation upon which to integrate the other reliability services. The benchmark study will help you fill in gaps by identifying existing internal best practices and techniques to yield the desired results. It will also help define a set of reliability practices to move defect prevention and detection tasks as far upstream in the development cycle as possible. Once you complete the assessment, choose the specific software techniques and integrate Software Reliability throughout the lifecycle of your program.

In the next several sections, I show the different Software Reliability techniques including which life-cycle phases I recommend.

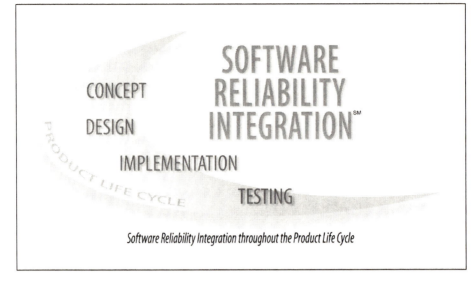

Software Reliability Integration throughout the Product Life Cycle

Figure 31.1: Software Reliability Integration

31.3.2 Software Reliability Integration in the Concept Phase

In the concept phase, there are two main Software Reliability techniques:

1. Software Reliability Goal Setting
2. Software Reliability Program Plan

Perform a Software Reliability Goal Setting by defining system-level software reliability goals. These goals become part of the overall Software Reliability Program Plan. Apply the goals to the design and testing phases.

31.3.3 Software Reliability Integration in the Design Phase

In the design phase, there are six main Software Reliability techniques:

1. Facilitation of Team Design Template Reviews
2. Facilitation of Team Design Reviews
3. Software Failure Modes and Effects Analysis (SFMEA)
4. Software Fault Tree Analysis (SFTA)
5. Software Failure Analysis
6. Software Fault Tolerance

Use the technique Facilitation of Team Design Template Reviews to conduct group pre-design review meetings, which provide your team with forums to expand their knowledge base of design techniques by exchanging design templates. Your team will greatly improve their design inspection results if the inspections are preceded by brief, informal reviews that are highly interactive at multiple points throughout the progression from system architecture through low-level design. This is known as the Facilitation of Team Design Reviews.

Use Software Failure Modes and Effects Analysis (SFMEA) and Software Fault Tree Analysis (SFTA) to identify and mitigate failure modes in software, similarly to how you used FMEA and FTA for hardware.

Prior to the final stage of a design, use Software Failure Analysis to identify core and vulnerable sections of the software that may benefit from additional run-time protection by incorporating Software Fault Tolerance techniques. See Section 39.4 on Software RCA for more details on Software Failure Analysis.

31.3.4 Software Reliability Integration in the Implementation Phase

In the implementation phase, there are two main Software Reliability techniques:

1. Facilitation of Code Reviews
2. Software Robustness and Coverage Testing

For Facilitation of Code Reviews, use reliability reviews to target the core and vulnerable sections of code to allow the owner of the source code to develop sufficient synergy with a small team of developers in finding defects. Use system testing efforts to focus on efficient detection of software faults using Software Robustness and Coverage Testing techniques for thorough module-level testing.

31.3.5 Software Reliability Integration in the Testing Phase

In the testing phase, there are four main Software Reliability techniques:

1. Software Reliability Measurements and Metrics
2. Usage Profile-Based Testing
3. Software Reliability Estimation
4. Software Reliability Demonstration Tests

Use Software Reliability Measurements and Metrics to track the number of remaining software defects, to calculate the Software mean time to failure (MTTF), and to anticipate when the software is ready for deployment. You will be able to apply Usage Profile-Based Testing methods to emphasize test cases based on their anticipated frequency of execution in the field.

One important new technique in Software Reliability is Software Reliability Growth. I have included a discussion of this in Appendix A of this book. Appendix A also contains a discussion on Software Reliability Estimation and Software Demonstration Tests.

RELIABILITY INTEGRATION: Integrating Software and Hardware Reliability
I stated early in this book that your customers don't care if the failure they experienced is due to software or hardware—to your customer, a failure is a failure. Therefore, the software and hardware teams should work together during Goal Setting and when developing metrics and measuring the progress of their respective programs. This will dramatically cut down on the amount of "finger-pointing" when failures do occur later in the program.

CASE STUDY: Software FMEA for a Semiconductor Robot
For a semiconductor robotics company, the first step the user must do is "teach" the robot the different steps to perform for normal operation. While performing a Software FMEA, we determined a failure mode in which the user could input an out-of-bounds condition for the speed. If this occurs, then it is possible the robot would travel at a higher than desirable speed and potentially crash into a wafer stack. Crashing into a wafer stack had the highest non-safety related risk priority number (RPN) because of the potential cost of failure. As part of the Software FMEA process, our client mitigated this failure mode by putting out-of-bound conditions in the software to protect against this error condition. Using the Software FMEA technique, we were able to improve the reliability of the software in the design phase successfully.

Part IV
Prototype Phase

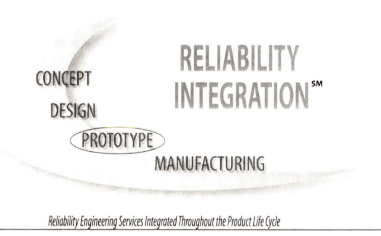

RELIABILITY
INTEGRATION[SM]

CONCEPT
DESIGN
PROTOTYPE
MANUFACTURING

Reliability Engineering Services Integrated Throughout the Product Life Cycle

32 Write a Comprehensive Test Plan—Don't Just Test to Specs

A **Reliability Test Plan (RTP)** is the high-level plan that calls out all of the reliability testing that you will perform on a product. Some of these tests are: Design Verification Tests (DVT), Highly Accelerated Life Tests (HALT), Reliability Demonstration Tests (RDT), Accelerated Life Tests (ALT), and Ongoing Reliability Tests (ORT).

With all of the reliability testing that goes on and all of the choices that you have for your testing, it is no wonder that many test programs wind up being fragmented and disjointed. The objectives of an RTP include the following:

1. To tie all of the different reliability test activities together
2. To ensure that you satisfy all of the testing needs
    a. regulatory testing needs
    b. testing to mitigate failure modes discovered during FMEA and other reliability analysis techniques used during the design phase.
3. To ensure there is no duplication between tests.

There is tremendous value in developing a comprehensive RTP that outlines all of the reliability testing activities. This can aid you in comparing with marketing specifications, can help you to satisfy your customers' requirements and expectations, and can help your test lab(s) perform the tests. RTP's often point out overlooked areas such as how to test a product, how to create a fixture, what constitutes a failure, and many other contingency plans necessary for successful test activities.

Prior to developing an RTP, perform an FMEA. As part of this exercise, determine what industry category your product will be classified as, what industry standards are required, and then what tests will be required at that point. The test plan is a living document. You may need to change it as you obtain results from your reliability analysis and as you start collecting test data (and possibly as a result of field data you collect from a previous generation of product).

RELIABILITY INTEGRATION: Integrating an RTP into a Reliability Program—Tailoring a Test Plan from Industry Standards
If you have a list of industry standards you need to comply with, then you should find the standard(s) most applicable to your product and use them as a baseline for your test plan. Then, tailor the plan to your specific industry and your specific application. Use the FMEA data to determine how you need to test specific components based on component failure mechanisms.

CASE STUDY: Your Product Environment is More than Its Use Environment
We were working with the National Science Foundation (NSF) developing a Neutrino Telescope, which is a three-dimensional array of electronic detectors designed to detect deep-space neutrinos that could provide evidence about the origins of the universe. This was a product whose end-use environment was to be buried two kilometers deep in the ice in Antarctica. This is an ideal environment for electronics for reducing external effects. The temperature is—40°C and is expected to vary by less than 1°C until the next ice age. You may ask, why would we perform any environmental testing at all for such a perfectly benign environment? The answer is in all of the stresses the product must go through to get to its end environment.

Here are the steps involved in manufacturing and transporting the product. For each step, I have identified the corresponding stresses:

1. Components are wave-soldered during manufacturing.
   Stress: temperature cycling
2. Boards are driven to integrator to put together into the end product.
   Stresses: temperature cycling, vibration
3. End product is put on a truck and driven to a boat dock.
   Stresses: temperature cycling, vibration, humidity
4. A boat takes the product from California to New Zealand.
   Stresses: temperature cycling, vibration, humidity
5. The product is then moved to an ice breaker ship.
   Stresses: cold temperature, vibration, shock

CASE STUDY: Your Product Environment is More than Its Use Environment, continued

6. The product is then put on a helicopter and flown to the installation site.
   Stresses: cold temperature, vibration

7. The product is then put on a snow cat and moved across the ice to its final installation point.
   Stresses: cold temperature, vibration

8. The product is then placed at the end of a drill and is buried in the ice using a process of melting and drilling.
   Stresses: vibration, shock, temperature cycling, rotational acceleration

You can see that the end-use environment was the least of our client's worries. Our client was worried that all of the environmental conditions they exposed the product to during transportation would weaken the product to the point that, when they finally lowered it into the ice, it would have very little life left in it. Also, if the product fails while in the ice (two kilometers deep in an ice shelf), it is nearly impossible to retrieve until the polar ice caps melt. This is because trying to re-drill the same hole to retrieve the product after positioning it would likely result in the drill hitting and destroying the product.

33 You Can't Start Testing Too Early

Almost every industry is competing on reliability, time to market, and product costs. Therefore, there is a need to develop more reliable products faster and at lower costs. Conventional reliability testing and improvement are often performed when development of a product is nearly complete, when time is short, and when any further improvements are more difficult and costly. Instead, **Early Reliability Testing (ERT)** is a development strategy that can provide higher reliability, with less cost and time for development, and less development risk.

Successive generations of a product are usually strongly correlated to each other in defects, wear, fatigue, failures, mechanisms, and root causes.

Arguments you will probably hear against early testing:

1. Immature Designs
2. Concurrent/Parallel Designs
3. Low Test Coverage
4. Few Samples
5. Immature Manufacturing Processes

However, you can address and overcome all of these issues, and when you do, you will reap huge benefits from ERT.

33.1 Immature Designs

Early in the product development process, designs are immature. Use HALT for early discovery of qualitative design defects. This will accelerate design maturation. The goal of HALT is to uncover design weaknesses (learn about qualitative issues) and expand design margins rather than to "pass" requirements.

The earlier you uncover and resolve defects, the more time and money you will save. This savings can be used for later quantitative tests, such as life testing. In particular, avoid conducting life tests on early prototypes because these prototypes are typically *not* built with final design materials or to a production intent process. For this reason, prototype defects, wear, fatigue, and failure may not be relevant to later generations.

33.2 Concurrent/Parallel Designs

Parallel or concurrent design and development requires connecting together two or more assemblies as a prerequisite to obtain meaningful test data. This inhibits the ability to perform functional testing prior to integration of the subassemblies. When this is the case, you can test these assemblies non-operationally under temperature and vibration to find resonant frequencies. This may point out many issues, including component interference issues, mounting issues, and board layout issues. Make sure to test under temperature and vibration because resonant frequencies change as a function of temperature.

33.3 Low Test Coverage

Early in the product development process, you typically have low in-house test coverage. Therefore you should test for qualitative (gross) issues (such as whether major functions and interfaces work properly), and postpone testing for quantitative (fine) issues (such as bit error rates during transmission). This early testing may be very worthwhile by saving considerable time and cost compared to later engineering redesign.

Occasionally, you can start with commercially available test equipment. This temporarily bypasses custom test programs and scripts that won't become available until later in the test phase. This commercially available test equipment is often good enough for worthwhile early testing.

33.4 Few Samples

Typically, only a few samples are available early in the product development process. Therefore, engineers previously avoided early testing. Instead, from even a few samples (sometimes just one), you can use a test/analyze/fix method and gain early qualitative feedback.

During later development with numerous samples, you should test for quantitative (fine) defects and failures to gain feedback in order to prove the length of the product's expected life.

There is useful synergy between these two tests. Typically, samples will fail within a fairly tight distribution. Therefore, you can use HALT to trade test margins for a size of specimen population. Thus, even with only a few samples, performing HALT at the outer edge of this distribution will tell you about the product robustness.

33.5 Immature Manufacturing Processes

Early in the product development process, samples were likely made with immature manufacturing processes. Augment the testing with failure analysis to determine the root cause of the failure. Often you can identify the cause of these failures using a simple examination (e.g., naked eye or simple microscopic examination). Performing RCA will enable you to:

1. Exclude failures probably restricted to immature manufacturing. These are nonrelevant failures.
2. Include failures probably significant for mature manufacturing. These are relevant failures that are either:
   a. not related to the manufacturing processes
   b. related to the manufacturing processes but are part of the processes that are already mature.
3. Distinguish probably relevant failures versus probably nonrelevant failures.

Relevant failures discovered early may be Golden Nuggets. These may forewarn what could happen with a mature process.

RELIABILITY INTEGRATION: Integrating Testing Across the Product Life Cycle—Knowledge from Early ERT Can Help During Later Testing
One argument against early testing is that the product will change, and therefore the failure mechanisms will change. While it is true that your product is likely to change, the failure mechanism and the testing methods by which you can find and prove out these mechanisms may in fact be similar. Therefore, early testing can really pay off because you can develop an understanding of how you will test the product later in the program. An example of this is with the solid state memory industry when they prove out the life of the memory cells. The failures you will likely be looking for are hot carrier injection and dielectric breakdown. You can start testing early and run some trial tests well before the product is ready for life testing to prove that your test methods and test fixture could indeed find these types of failures.

CASE STUDY: Using ERT for a Glucose Meter

We were working with a medical company making a glucose meter. The meter had the potential of revolutionizing the market because of its small size and increased functionality. Due to the size constraints and tighter tolerances that go along with a smaller product, it was quite a challenge for the mechanical engineering department. We convinced them that they needed to get some systems (or possibly even just the mechanical portion of the systems) on test very early, well before life testing, in order to perform characterization and repeatability studies. We set up a high-speed camera and took pictures at several points where the glucose strip went through the meter, then placed reference markers on the camera to measure how repeatable the process was. As a result of the testing, we found that there was enough variation from run to run of the strip that the strip could jam if the tolerances stacked up, causing out of tolerance conditions. With this data, our client was able to make modifications in the amount of variation allowable to a few key dimensions of the mechanical portion of the product, and they were able to do this very early in the development program, well before they integrated the mechanical portion with the rest of the meter.

HALT, Who Goes There?

Highly Accelerated Life Test (HALT) is a design technique that you can use to discover product weaknesses and design margins. Throughout the HALT process the intent is to subject the product to stimuli well beyond the expected field environments in order to determine the operating and destruct limits of your product. Failures that typically show up in the field over a long period of time at much lower stress levels are quickly discovered while applying high stress conditions over a short period of time. In order to "ruggedize" your product, you need to determine the root cause of each failure and provide corrective action and then continue stressing further, repeating this process until you reach the Fundamental Limit of the Technology (FLT) for your product. The definition of FLT is the limit at which you can't go any higher with a stress without changing the failure mechanism and producing nonrelevant failures. This process will yield the widest possible margin between product capabilities and the environment in which it will operate, thus increasing your product's reliability by reducing the number of field returns.

You can also use the operating and destruct limits discovered during HALT to develop an effective Highly Accelerated Stress Screen (HASS) for manufacturing, which will quickly detect any process flaws without taking significant life out of your product. The HASS process can ensure that you maintain the reliability gains achieved through HALT in future production.

Typical stresses used during HALT are cold step stress, hot step stress, rapid temperature transitions, vibration step stress, and combined thermal and vibration environments.

Note that these are typical stresses, but you can use other stresses as well. HALT really means anything you can think of to accelerate the testing process to find design weaknesses before your customers find them. What are some other methods for HALT? Could electrostatic discharge (ESD) testing utilize the HALT process? Why not? Couldn't you continually increase the amount of electrostatic voltage to the specification point, then gradually increase the voltage until the product fails? Sure you could. You could probably use the HALT process for drop testing, altitude testing, voltage margining, frequency margining, and just about any other environmental or electrical or mechanical stresses. I said "probably" because it ultimately depends on the failure mechanisms of the technologies in the product and what stresses can accelerate them.

How high can you go with stressing your product? You should stress your product until you reach its FLT. This means that after you reach an operational or destruct limit, you should create workarounds to get by that failure and move to the next failure and follow this process until you reach your product's FLT. Most products have more than one FLT—in fact, products have as many FLTs as there are unique technology types. Even after reaching an FLT for one type of technology, remove that component or assembly which contains that technology and then continue on to search for the next design weakness and do this until you reach the next FLT. Continue this process until you can't go any further (usually because you can't easily remove the weak assembly from the stress).

What failures should you fix? That is a hotly debated topic among HALT practitioners. You should analyze all failures and fix the ones that are more likely to occur in the field. One rule of thumb is "If it is inexpensive to fix, fix it. If not, then analyze." This could save your engineering team a lot of debate over which failures to fix. However, there is another method that you can use. You can determine some technological limits even before testing. You can find these during the FMEA process. If you can determine during the FMEA whether a failure mode would develop into a relevant failure or not, you could save a lot of test time.

Let's say, for instance, you are using a certain type of plastic in your product and you are about to put your product through the HALT process. Wouldn't it make sense to determine the limitations of that plastic *before* you put it through HALT? You may not want to destroy the plastic casing "just to see what will happen." Worse, what if the plastic casing warps enough to cause a secondary, more catastrophic failure that would never have happened if you hadn't overstressed the plastic in the first place? Good engineering judgment would tell you to remove the plastic from the test *before* you reach the FLT of the plastic. This will save you a lot of time and money. If the plastic is required for the mechanical integrity of your product, then put the plastic back on during the vibration stress. Some would argue that the plastic is also required for the thermal testing, and without it the airflow will be completely different. This is likely to be the case, but you shouldn't care because you aren't running a qualification test but rather a discovery process. You want to know how the product fails, and you aren't as concerned with the exact temperature at which it occurred. Removing the plastic will allow faster temperature cycling to accelerate the weak areas to failure.

You can predetermine some product limitations this way, but some you can't. Let's not mistake an FLT with a component specification. A piece of plastic has a known FLT because the phase transition point of the plastic from solid to liquid is well known and well documented. This isn't true for an integrated circuit (IC) or other electronic components. If you look at the specification sheet of an IC, you will only find the specification limits or what the manufacturer guarantees for the product, not the FLT. This is important to understand when planning a HALT so that you don't stop testing too soon. Performing research before a HALT is a key part to planning out the HALT.

The reason HALT works is because it follows the Stress versus Number of Cycles (S-N) relationship. See Figure 34.1 for a diagram of this relationship. If your product is exposed to N0 number of cycles of field stress S0, then it will fail. In HALT, we increase the stress, thereby decreasing the number of cycles to failure. This relationship holds true for just about every component in your system. What varies from component to component is the slope of the line (and in many cases, the relationship won't even be linear). The other thing that holds true for each component is that there is a point in which you can no longer increase the stress without changing the failure mechanism. We call this the point of nonrelevant failures, or the FLT. This stress level also varies from component to component.

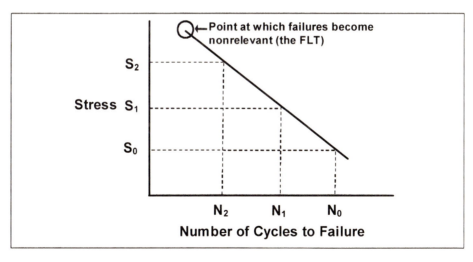

Figure 34.1: Why HALT Works—The S-N Diagram

To understand this concept better, let's look at a graph in Figure 34.2. The graph highlights temperature as the stress to illustrate the concepts of Product Operational Spec, Upper and Lower Operating Limits, and Upper and Lower Destruct Limits. The Product Operational Spec is the specification that you tell your customers your product can meet. An operational limit is one in which the failure recovers when you reduce or remove the stress, and a destruct limit is one in which the failure doesn't recover when you reduce or remove the stress.

In Figure 34.2, we have drawn a line for each of these points; we have also drawn distribution curves around each line. These are failure distributions, and we must draw them as distributions because not every product will fail at the same stress point, but rather they will fail within a distribution. Note that we have drawn the distributions as normal or Gaussian distributions (evenly distributed about a center point) just for the sake of illustration, but they may not be normal at all. We don't have enough samples to determine actually what type of distribution, the shape of the distribution, or even how long the tails are on each side, so we are drawing this for illustration purposes. The distribution we drew around the Product Operational Spec lines is the distribution of use by your customers. You will have customers that take your product outside of your specifications, and this distribution captures those situations. Notice in Figure 34.2 that there are failure opportunities where the Lower Operating Limit upper tail intersects with the Product Operational Spec lower tail, as well as where the Upper Operating Limit lower tail intersects with the Product Operational Spec upper tail.

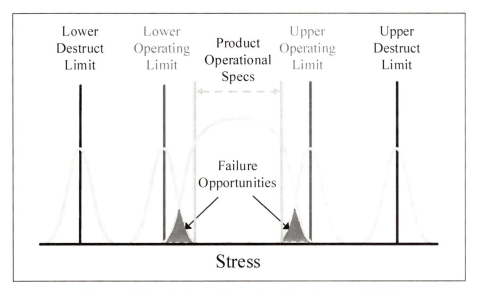

Figure 34.2: Failure Opportunities before HALT

We use the HALT process to discover first what the limits are and then to expand the limits, thereby reducing or removing these failure opportunities. In Figure 34.3, we have expanded the operating and destruct margins, and in the process, we have removed the failure opportunities. We already stated that we don't really know what the distributions look like. Therefore, we will never know if we have completed removed these failure opportunities. That is why we don't stop at any preset stress points in HALT but rather continue stressing until we reach the product's FLT.

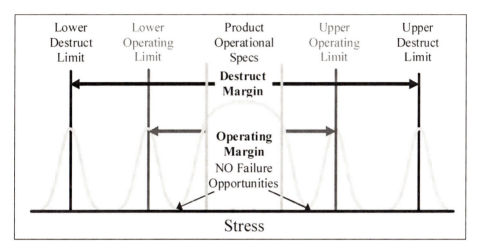

Figure 34.3: Removing Failure Opportunities through HALT

34.1 HALT Process

There are 11 main steps to the HALT Process:

1. Perform Failure Modes and Effects Analysis (FMEA)
2. Write HALT Plan
3. Prepare for HALT
4. Choose a Location to Perform the HALT
5. Develop the HALT Fixture
6. Set Up the HALT
7. Perform HALT
8. Write a HALT Report
9. Perform Corrective Actions
10. Perform a Verification HALT
11. Perform Re-HALTs and Periodic HALTs

34.1.1 Perform Failure Modes and Effects Analysis (FMEA)

The first step is to perform the FMEA, which we have discussed in detail in Chapter 16. During the FMEA, determine the highest risk areas of your product. As part of the mitigation process for the FMEA, you have two choices:

1. Come up with corrective actions for these risks and identify tests to prove that your corrective actions have sufficiently mitigated the risk.
2. Determine that a particular risk may not be as significant as you originally thought. You may need to come up with a test to prove this.

In both cases, these are inputs to your testing program, and more specifically, they may be inputs to your HALT Plan if you determine HALT is the best technique to prove the mitigation for the particular failure mode.

34.1.2 Write HALT Plan

Next, write the HALT Plan. In this plan, you should determine the types of stresses, levels of stresses, and order of stresses. You should also determine the number of samples, functional tests, and parameters to monitor, and determine what constitutes a failure. Decisions in this plan will dictate the relative success of the HALT. Writing a HALT Plan is probably one of the least understood portions of HALT. Many people just take standard plan templates you can download from the internet and use them with little regard for the specific attributes of the product itself. However, every product is unique, and therefore you should tailor the HALT Plan to fit your product. You should use the FMEA process to understand better the components that have limitations and determine what these limitations are *before* you write the HALT Plan. As part of the FMEA, you may need to perform additional research on a specific component, or you may need to call a component manufacturer to review their reliability test data. If you don't get sufficient information from these two methods, you may want to perform component level testing on the suspect component types before performing HALT at the system level. This will save you a lot of time and money if you don't have to discover something in testing that you could have figured out through analysis.

34.1.3 Prepare for HALT

Whether you are new to the HALT process or a veteran, there are many steps that you can take to help prepare for the HALT to make it run more smoothly. These steps include alerting the appropriate experts to be available to trouble-shoot problems if needed, having spares available, making certain the appropriate test equipment is in place, and developing workarounds in case failures occur (including extension cables to route weaker assemblies outside of the chamber when they fail). In addition, we recommend that you perform a dry-run of the test routines prior to setting up at the HALT facility. We have experienced quite a few HALTs in which our client didn't perform a dry-run of the test routines prior to the HALT and discovered during the HALT that the routines weren't working properly, resulting in us troubleshooting failures that were not real. In each case, this cost us several days in test time. All these preparations will minimize the amount of downtime during the HALT process.

34.1.4 Choose a Location to Perform the HALT

If you have a HALT chamber in-house, this is the most desirable situation because your engineers can be more involved in the process. Just make sure you understand the concepts described in this chapter. If you don't have a chamber in-house, then choose your test lab wisely. There are obvious advantages for the lab to be close to your facility for the convenience of your engineering staff, but make sure the lab understands the concepts of HALT. In 1995, I started a HALT lab in Santa Clara, California, as part of Qualmark Corporation. At that time it was only one of a handful of HALT labs in the world. Over the past fifteen years, dozens of HALT labs have sprung up around the world, and some are better than others in terms of their understanding of the HALT process and their ability to guide you through it.

Make sure to ask the lab for a sample HALT Plan and HALT Report, and study these to make sure that the personnel at the lab considered the uniqueness of the product that they subjected to HALT and accounted for this during the HALT. In addition, make sure that the lab performed HALT on the product to its failure limit, and make sure the lab troubleshot failures to provide workarounds. Here are some guidelines on what to look for when choosing a lab facility:

1. The HALT should be run by a degreed engineer with at least five years of experience in HALT.
2. The lab should have peripheral equipment—thermocouples, accelerometers, a signal analyzer, and a data logger at a minimum.
3. The lab should have failure analysis equipment or have a partnership with a failure analysis lab nearby to be able to perform root cause on failures that occur during HALT.
4. If possible, the lab should be within driving distance in case you need to go back to your facility for spare parts or for additional help.

34.1.5 Develop the HALT Fixture

The HALT fixture typically is comprised of a thermal airflow scheme as well as a vibration fixture. The thermal airflow scheme is designed to maximize airflow across the product. This consists of directing air ducts on one side of the product and having the air blow across the product. It is usually best to take the cover off the product to improve the airflow. However, you may face situations in which you can't remove the covers, either for structural reasons or for issues of EMI susceptibility (with the covers to your product off, outside EMI noise may effect the product during the test). In these situations, you should direct the airflow into the openings of the product and allow for the air to circulate through the product and blow out the other side. Never blow air into the input and exhaust of the product simultaneously. Always blow the air into the direction of the natural air path or direction the fans are pushing the air (if the product has fans). Attach thermocouples to various places on the product and in the air duct and take measurements of the ramp rate differences between what the product experiences versus the chamber ramp rate. The key is to get the product ramp rate as close to the chamber ramp rate. The faster the rate of change, the faster you will discover potential product weaknesses.

For vibration, design a fixture to maximize the vibration transmissibility from the vibration table to the product itself. This usually requires using aluminum bars or channels above and/or below the product to secure it to the table. Attach accelerometers to both the product and the vibration table. The key is to get the product vibration level as high as possible with respect to the table vibration. We recommend you use a signal analyzer for this measurement so that you can measure and compare table versus product vibration for both the total amount of vibration as well as for each band of frequency.

34.1.6 Set Up the HALT

Setting up a HALT may take a few minutes or a few days, depending on the complexity of your product. Make sure you have all of the necessary equipment lined up and available for the testing. Having two cables available for every connection will save you a great deal of time during the troubleshooting process. Make sure you have the ability to troubleshoot failures and have spare components or assemblies available in case failures occur. Make sure you have extender cables ready in case you need to separate out assemblies as failures occur. I've seen tests delayed by a few days due to an extender cable not built ahead of time. Apply thermocouples and/or accelerometers in accordance with your HALT Plan. Make sure that your engineering staff is available and at the site (or on call if they can't be at the site).

34.1.7 Perform HALT

The actual HALT itself consists of the following steps:

1. Follow the HALT Plan and start with the first stress.
2. Increase the stress until a failure occurs. This can be an operating limit, or it can be a destruct limit.

3. Review the failure to determine if it is relevant. Go back to your FMEA to determine if you predicted this failure and what you determined about the relevancy at the point of failure. If you didn't perform an FMEA, or if your FMEA didn't predict this failure, then you will need to perform a Root Cause Analysis (RCA) on the failure to determine its relevancy. See Section 34.1.9 for a discussion on RCA during HALT. The failure isn't relevant if it occurred outside the product's FLT.

4. Provide workarounds to get past a failure and then continue stressing the product to find the next relevant failure. This may entail removing portions of the product with extender cables, and it may entail disabling protection circuitry. This is one area where there is no substitute for experience when it comes to the HALT practitioner.

5. Continue this process until you reach the FLT. Remember, your product probably has more than one FLT so don't stop until you have reached as many FLTs as you can, and don't stop until you aren't able to separate the assemblies any further. See Figure 34.4 for a diagram of performing HALT to the FLT.

6. Choose the next type of stress and follow the same process.

7. Combine the different types of stresses and follow the same process.

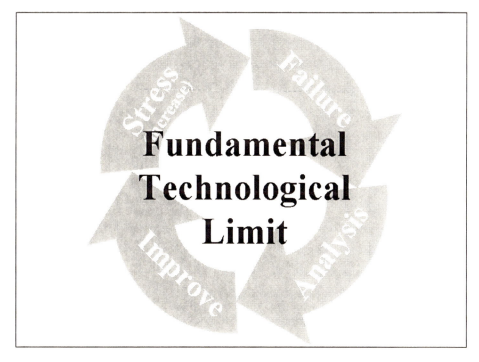

Figure 34.4: Performing HALT to the FLT

34.1.8 Write a HALT Report

In your HALT Report, make sure to document all of your findings, including at what point each failure occurred and what steps you took to provide a workaround for each failure. Pictures, charts, and graphs are great for your management. However, the most important parts of the report are your recommendations and the follow-up actions you plan to take.

34.1.9 Perform Corrective Actions

Log all relevant failures into your Closed Loop Corrective Action (CLCA) System. Provide corrective actions for each of these failures. For some of the corrective actions, you may have determined a suitable solution during the HALT, so your workaround may be a perfectly suitable corrective action that you can implement (such as applying RTV or other silicone material to ensure that a component doesn't break off). However, some workarounds aren't suitable corrective actions. They were suitable for allowing you to continue the HALT process, but they may not be acceptable for production (such as using tie-wraps to hold connectors in place), and you should come up with another means that is more suitable for your product. For some failures, you may need to perform a detailed RCA process to determine the cause of the failure and its relevancy. There may even be instances where you shouldn't provide corrective action for some relevant failures. This is usually due to the high cost of the corrective action or the low probability of the failure occurring, or both.

34.1.10 Perform a Verification HALT

After you perform HALT and provide corrective action for a specific failure, then perform HALT again to ensure that the corrective action improved the product performance and did not introduce new failure modes. This second HALT is called a Verification HALT. By conducting a Verification HALT you will ensure that your corrective action improved the product margins. Make sure to subject the product to all of the stresses you used for the HALT that uncovered the failure because frequently corrective actions may improve the product margins for one stress but reduce the product margins for another stress. For example, if you have a failure in vibration caused by board flexing, and as a corrective action you add a board stiffener, you may improve the vibration performance of the product, but you may affect the thermal airflow if the board stiffener inhibits the airflow across the product.

Note that during Verification HALT you can skip the lower stress levels for each of the stresses and jump to within one to two steps of the original failure point. For example, if the original test plan called for increasing the temperature in 10°C steps starting at 20°C, and if a failure occurred at 80°C, then during the Verification HALT, start at 60°C or 70°C.

You may even need to use additional stresses beyond the one you used in your previous HALT if the change involves a new technology.

34.1.11 Perform Re-HALTs and Periodic HALTs

Re-HALT is the process of performing HALT later in the development process after the product has matured, more samples are available, and test routines are more complete. We recommend you use HALT as a review step in your engineering change order (ECO) system; a representative from your design team or your HALT team should determine if a Re-HALT is required before you implement the change. Usually HALT is only required if the change is major, such as a new technology or a change in a circuit. For minor changes such as adding vendors to your approved vendor list (AVL), you may not need to perform a Re-HALT. I recommend having a representative from your ECO review board responsible for this decision because you will need to use engineering judgment to make this decision. If your company can't add this representative to your ECO review board, then I recommend you perform Periodic HALT regardless of the changes you have made to your product. This is due in part because your components may have performance drifts due to batch variations, or worse, your component vendors may change their component processes without notifying you. A good rule of thumb is to perform Periodic HALT once a quarter during active production of any product. Of course, the volume and product cost will play a big role in setting this time period.

For Re-HALT and Periodic HALT, you may skip the lower stress levels for each of the stresses and jump to within one to two steps of the original failure point. You may even need to use additional stresses beyond the one you used in your previous HALT if the change involves a new technology.

34.2 Summary of HALT Results at an Accelerated Reliability Test Center

Appendix B provides backup data for HALT in the form of a study that I performed based on my experience running a HALT lab from 1995–2000. I have reviewed more recent data since this time, and although some of the data has changed, the results remained largely the same. The main reason I performed the study and the reason I am including it in this book is because engineers often approach me and tell me that they understand the concepts of HALT but are worried that the methods won't work for their particular product. Therefore, by collecting a large amount of data and demonstrating the vast number of industries we have successfully implemented HALT, I have been able to convince engineers over the years that HALT, when applied correctly, is a very valuable technique.

RELIABILITY INTEGRATION: Integrating Design Techniques with HALT
HALT is usually the first technique you turn to when proving the reliability of your product through testing. However, in order for your testing to make sense, you need to know what you are looking for, and this is where you can get help from all of the design techniques you used. I describe a variety of design techniques in Part 3 of this book. Techniques such as FMEA, Predictions, Derating Analysis, Thermal Analysis, and others all provide valuable input to the HALT Plan. This is an area where I find some HALT labs lacking in skill, as they don't tailor their approach when performing HALT on a product. If you download a free generic HALT plan template from the internet, you will get exactly what you paid for it—nothing.

CASE STUDY: HALT for a Smart Card
We were working with a smart card company developing a HALT Plan for their new smart card credit card. Inside the card was a microprocessor for storing the information rather than using a magnetic strip. During the FMEA, we brainstormed different ways the card would be used in order to determine how the card could fail. This revealed how our client would need to test the card and how we could accelerate the testing. We determined that one of the dominant failure mechanisms was that the microprocessor could crack or the leads to the microprocessor could break due to bending (possibly from putting the card in a wallet and then sitting on it). We also came up with spill scenarios as well as being damaged from poking it with sharp objects (possibly from being inside of a pocket or a purse). Clearly the traditional temperature/vibration we associate with HALT wasn't going to work here. For one of the tests, we developed a stress scenario composed of exposing the card to increased bending until we achieved the ultimate breaking point. Can this be considered HALT? To answer this, let's look back at the definition of HALT—"HALT is a design technique that is used to discover product weaknesses and design margins. Throughout the HALT process, the intent is to subject the product to stimuli well beyond the expected field environments to determine the operating and destruct limits of your product." Based on this definition, if we set up the test to determine product weaknesses and design margins then we are performing HALT— regardless of the type of the stress.

35 Demonstrate Your Reliability

A **Reliability Demonstration Test (RDT)** is the process of demonstrating the steady state reliability of a product throughout testing. An RDT is usually performed at the system level and is typically set up as a success test. If you need to understand the estimated steady state reliability of a product prior to shipping the product to your customers, you can use an RDT to determine its MTBF. Reliability Predictions are valuable early in a design cycle, but later in the design cycle, an RDT can give a more accurate estimate of reliability.

Using an RDT, you will be able to approximate the mean time between failure (MTBF) within a desired level of confidence. This shouldn't be confused with an Accelerated Life Test (ALT), which is used to determine when a product or portion of a product will reach its end of life (when the product will wear out). The RDT is typically performed in an accelerated manner using both environmental and electrical stresses so you can determine the MTBF before you start shipping the product. However, you are only accelerating the testing within the steady state portion of the product life cycle. The most common environmental accelerant is high temperature, and the most common electrical accelerant is power cycling. Figure 35.1 shows the Reliability "Bathtub" Curve, highlighting where RDT is most effective. Note that RDT is useful for demonstrating reliability only within the steady state region. One advantage of using RDT in the steady state region is that you can assume all failures are randomly distributed. When this is the case, you can substitute samples for test time to accelerate the testing further.

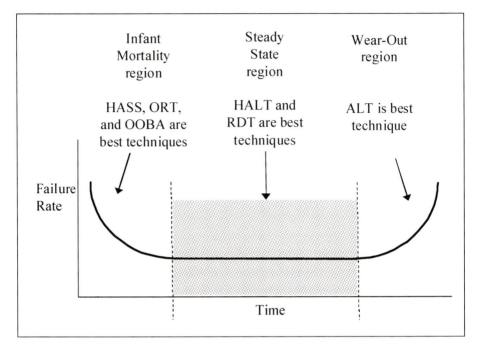

Figure 35.1: RDT Region of Reliability "Bathtub" Curve

There are four stages to an RDT:

1. Planning
2. Testing and Monitoring
3. Data Analysis
4. Report Writing

35.1 Planning

In the planning stage, first establish a Reliability Test Goal (you can often get this directly from the Reliability Prediction). Then, if you plan on accelerating the test, choose an acceleration model (such as the Arrhenius Model if you choose high temperature as your accelerant). Next, develop an RDT Decision Matrix and outline all of the parameters prior to writing the plan, including types of stresses, number of samples, length of test, and confidence level, along with advantages and disadvantages for each. From this matrix, decide on all of the parameters that will go into the RDT Plan.

Most RDTs are set up either as testing for a pre-determined amount of time or until a pre-determined number of failures occur. This is often referred to as a Probability Ratio Sequential Test (PRST). PRSTs are based on the ratio of an acceptable MTBF (which should have a high probability of acceptance) to an unacceptable MTBF (which should have a low probability of acceptance). I have shown a sample PRST plot in Figure 35.2.

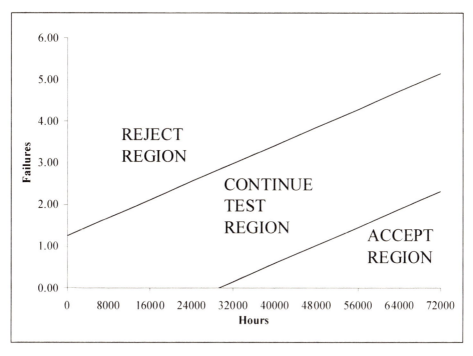

Figure 35.2: Sample PRST Plot

The plot illustrates three distinct regions—the reject region, the continue test region, and the accept region. Periodically, during the testing, calculate the results achieved to date. Based on where it falls on the plot, either stop and accept the samples (this indicates that the samples will meet the MTBF target), reject the samples (this indicates that the samples will *not* meet the MTBF target), or continue testing (this indicates that you need more test data to determine if the samples will meet the MTBF target).

35.2 Testing and Monitoring

In the testing and monitoring stage, set up and start the RDT and monitor the results for failures. You should set up the test so that you will only need to monitor a few times a week, once a day at most. Automatic monitoring is even better to get a more accurate time of failure. If failures do occur (I say "if" because often we have seen RDTs run with no failures at all), then you need to perform Root Cause Analysis (RCA) and log the failure into your Closed Loop Corrective Action (CLCA) System. Depending on the nature of the failure and corrective action, you have three choices:

1. Stop the test and replace all the samples under test with samples containing the corrective action (in which case you must start the testing clock over relative to that specific failure mode).
2. Replace the one failed sample with a fixed sample.
3. Continue the testing with one less sample.

For some failures, you may determine that the failure wasn't relevant and shouldn't be counted if either of these two situations exists:

1. The failure was deemed to be an infant mortality. Remember that the goal of the test is to determine the MTBF in the steady state region of the product life. Therefore, you shouldn't include infant mortality failures. Make sure you perform a corrective action to determine conclusively that it was an infant mortality failure. Don't just use a time-based approach, such as assuming that all failures within the first 72 hours are automatically classified as infant mortality.
2. The failure was due to a technology limit for one area of the product that isn't of interest to you, or the stressing wore out one particular portion of the product, and this also wasn't of interest to you. As an example, we ran an RDT on a server product that had a hard drive in it. To get sufficient acceleration for the testing, we ran the test at 70°C, knowing that this would cause the hard drive to wear out prematurely. We understood this wear-out mechanism for the hard drive and weren't interested in it because we determined that 70°C was at or above the hard drive's FLT for long-term wear, and we never expected the product to experience this in its use environment. When the hard drive failed, we just replaced it and continued the testing, and we didn't count these failures when calculating the product MTBF.

35.3 Data Analysis

Periodically, during the testing, you should calculate the achieved MTBF at that point in the test and compare the results against your original PRST plot. By doing this, you can determine whether you should stop the testing and publish the results, or whether you should continue testing. Make sure that you provide a status to the team on the progress of the testing so that they have clear and updated expectations.

35.4 Report Writing

In the report, compare the achieved results to the goal and show how you calculated the MTBF. Make sure to describe each failure and what actions you took as a result of the failure.

As a side note on statistical confidence of a test, I was teaching the Certified Reliability Engineer (CRE) Preparation Course several years ago and came to the topic of confidence during the testing. We typically recommend a confidence level between eighty and ninety percent. This means that after you perform the test, you have an eighty to ninety percent chance that the sample population revealed the same results as the entire population will once you ship product into the field. I asked the students how much confidence they would like to have in their test results. One student said that his management required 100% confidence in all their testing. A 100% confidence would require that you test every single product produced to the end of life. That would give you nothing to ship because you will have every product in your test. It is impossible to get 100% confidence. By the very nature of sample testing, you are giving up some level of confidence.

RELIABILITY INTEGRATION: Integrating RDT with Predictions—RDT Using High Temperature as an Accelerant
When performing an RDT using high temperature as an accelerant, first perform HALT to find the high temperature limit. Then, as a guideline, back off by 10–20°C and set this as your RDT temperature. Next, calculate the acceleration factor by setting the Reliability Prediction usage environment to the RDT temperature. The acceleration factor is the factor that denotes how much you are accelerating the test beyond the end use conditions. Using the Arrhenius Model, determine the activation energy by using the revised temperature to predict the MTBF difference between the two temperatures. This ratio then becomes the acceleration factor. This is essentially a weighted average method of calculating the activation energy. Using this acceleration factor will enable you to demonstrate the reliability of your product in a much shorter period of time and have more accurate results.

CASE STUDY: RDT using Temperature Cycling as an Accelerant

We helped a power supply company perform an RDT for their new power supply. We asked our client how the product was going to be used. Its intended use environment was in a semiconductor fabrication plant, so it could be subjected to some high temperature but very little temperature cycling or vibration. Therefore, we recommended an RDT using high temperature as the accelerant. Our client thought that temperature cycling had a greater acceleration factor than high temperature and asked if we could use that instead. First of all, temperature cycling doesn't always have a greater acceleration factor than high temperature. It largely depends on the failure mechanism. Secondly, you can't use temperature cycling as a stress if the product doesn't experience temperature cycling in the field. You would have no way of calculating the acceleration factor and it is likely that the failure mechanism you discover would be different than the one that would have occurred in the field. We convinced our client to stay with high temperature only as the stress and ran a day RDT, which proved a five-year MTBF. This did answer the question of what our client hoped to achieve.

KEY DIFFERENCE BETWEEN HALT AND ALT: For RDT, you are limited to only using the stresses the product will experience in the field because in order to use the stress, you must be able to calculate an acceleration factor. For HALT, you aren't limited to only using the stresses that the product will experience in the field because in HALT, you are only interested in the weakness, not in how the weakness was uncovered. Case in point—you can use vibration to find a weak solder joint even if the product only experiences temperature cycling in the field.

36 Accelerated Life Test (ALT) for Those Life-Limiting Failures

An **Accelerated Life Test (ALT)** is the process of determining the useful life of a product in a short period of time by accelerating the use environment (how the product will be used). ALTs are also good for finding dominant failure mechanisms. ALTs are usually performed on individual components or assemblies rather than full systems. ALTs frequently are used when there is a wear-out mechanism involved, thus precluding you from directly substituting samples for test time as in an RDT.

Figure 36.1 shows the Reliability "Bathtub" Curve, highlighting where ALT is most effective. Note that ALT is useful for measuring reliability only within the wear-out region.

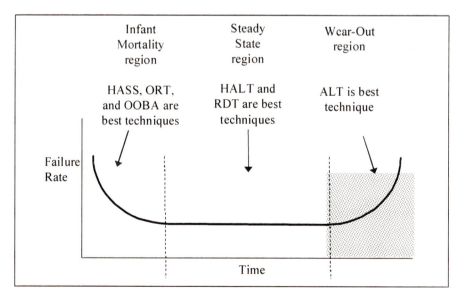

Figure 36.1: ALT Region of Reliability "Bathtub" Curve

If you need to understand the estimated life of key assemblies within your product prior to shipping, you can use an ALT to determine this. End-of-Life (EOL) Analysis is valuable early in a design cycle, but later in the design cycle, an ALT can provide you with a much more accurate estimate of reliability, especially in specific use environments. By using an ALT on an assembly, you will be able to approximate how long an assembly will last before it fails, as well as what the failure modes will be when it does fail. ALT is performed in an accelerated manner so you can find out this information as quickly as possible.

36.1 Value of ALT

ALTs have tremendous value in many areas:

1. Estimate the life of the product in the field.
2. Identify specific failure modes that will occur in the field before the product is shipped.
3. Allow you to prioritize product improvements.
4. Map product risks to various environments.
5. Determine if the product is ready to ship.
6. Determine when key components of the product will wear out.

36.2 Steps of an ALT Process

There are five stages to an ALT:

1. Strategy
2. Planning
3. Testing and Monitoring
4. Data Analysis
5. Report Writing

36.2.1 Strategy

In the strategy stage, determine which components or assemblies you will subject to the ALT. Unlike the RDT, which you typically perform at the system level, for ALT you perform at the component or assembly level because you are usually interested in specific failure mechanisms. Once you identify the component, determine the dominant failure mechanism and how to accelerate it. This may be quite well known, or it may require a bit of research. It may even require some experimentation.

A number of common acceleration models exist, such as the Arrhenius Model, the Coffin-Manson Model (a model used to describe mechanical fatigue in material and growth of cracks in solder and other metals due to repeated temperature cycling), and others. You will need to perform research to determine which model is the best to use. If you use the incorrect model, the entire test may be invalid. Once you determine the model you will use, then determine the proper coefficients for the model. If you can't determine this from your research, you will need to set up the ALT so that you determine the coefficients during the testing. For example, if you know that the failure mechanism can be accelerated by using high temperature, and you decide to use the Arrhenius Model, you still need to know the activation energy of the failure mechanism to use the model. If you don't know this, you can set up the ALT at several different levels of stress and, as long as you have at least one failure at each stress level, you can calculate the acceleration factor directly from the test results.

If you are uncertain whether the model you have chosen is correct, or whether you have chosen the correct values for the coefficients, consult an expert reliability firm.

After choosing a model, you need to determine the slope of the Weibull Distribution (β). This parameter β describes how fast the population will wear out once wear-out begins on the first sample. If wear-out is present, then β will be greater than 1 but how much greater? A β of 1.5 indicates a gradual wear-out, whereas a β of 4 indicates a sudden wear-out. There are various sources where you can determine the β factor. The first place to look is the internet. Your component vendor may also have this information. If you can't determine this information, you can proceed writing the test plan without it, but you won't have an accurate estimate for the length of the test.

36.2.2 Planning

In the planning stage, first establish a Reliability Test Goal for the test (often you can get this directly from the End-of-Life (EOL) Analysis). Develop a Decision Matrix and outline all of the parameters that you plan to use prior to writing the plan, including types of stresses, number of samples, length of test, confidence level, and Beta (β) (the slope of the Weibull Distribution), along with advantages and disadvantages for each. From this matrix, decide on all of the parameters that will go into the ALT Plan. If you haven't already determined the parameters for the model (such as the activation energy for the Arrhenius Model if that is the model you have chosen), you need to put this into your ALT Plan as well.

36.2.3 Testing and Monitoring

In the testing and monitoring stage, set up and start the ALT. Monitor the results for failures. The test should be set up so that you need to monitor a few times a week, once a day at most. Automatic monitoring is even better to get a more accurate time of failure. The only exception to this is if it will take too long to experience a failure, and at this point, you should set up the test to find a Leading Indicator rather than to test for a failure—see explanation of Leading Indicators in Section 36.3.

When failures occur, you may not need to perform a Root Cause Analysis (RCA) if the component fails in the manner you predicted during your EOL Analysis (but it is always a good idea to check rather than to assume).

36.2.4 Data Analysis

Periodically, during the testing, you should calculate the achieved life at that point in the test and compare the result against your original goal to determine whether you should stop the testing and publish the results, or if you should continue testing and for how much longer. Make sure you provide a status update to the team on the progress of the testing so that they have clear and updated expectations.

36.2.5 Report Writing

In the report, compare the achieved results to the goal and show how you calculated the life. Make sure to describe each failure and what you did as a result of the failure.

36.3 Leading Indicator ALT

There is a special type of ALT called a Leading Indicators ALT in which ALT is run not to the point of a failure but rather until you detect an indication that a parameter has changed. A good example of this is discussed in the case study in Chapter 25, Design of Experiments (DOE), in which we measured the acoustic and vibration noise output of a blower assembly to determine when it was starting to wear. A Leading Indicator ALT is an effective method of performing ALT for many different types of products—both mechanical and electronic. However, you may need to do a bit of research on the expected failure mechanism to determine what indications of failure to look for. The indications could be changes in temperature, vibration, radiated emissions, voltage, current, magnetic field, acoustic noise, weight gain/loss, power output, and many others.

36.4 Extrapolating Product Life Through Failure Analysis

Often you will not have enough time to test a product to failure, and you can't determine a leading indicator for which to monitor. When this is the case, determine if it is possible to extrapolate the life of the product through failure analysis based on the amount of wear that took place during the life test.

We subjected a linear motor to ALT, and rather than taking it to failure or using Leading Indicators, we subjected the motor to three months of ALT (equivalent to about five years of life in the field). We then sent the motors back to the manufacturer to analyze the amount of wear that took place. They estimated that the motors still had over half the life left in them. Therefore, we extrapolated that the motors would have survived out to six months of testing, which was equivalent to ten years of life in the field.

36.5 Accelerating the Duty Cycle

A product duty cycle is the percentage of time the product is operating in its use environment. Most products aren't used twenty-four hours a day, seven days a week. Even products that are used continuously may not be used in full load or full operation all of the time. When this is the case, if you are able to run the product at a higher percentage of time during the ALT, then you are accelerating the duty cycle.

Often, accelerating the duty cycle may be enough acceleration for your ALT. Sometimes you can get an acceleration factor of 100:1 (if the product is only used ten to fifteen minutes per day, such as for daily data backups). At other times, you may only get a 4:1 acceleration (if the product is used eight hours a day, five days a week, such as an office product). Sometimes, the duty cycle is only 2:1 or lower (if the product is used twelve hours a day every day, such as a solar product). Nevertheless, you can still use duty cycle as one of the accelerants for the ALT.

Take caution when using duty cycle for acceleration because it can be a bit tricky. Take for example a motor that turns on and off periodically. You may want to speed up the test by turning the motor on and off more frequently. However, you need to understand what is taking place when you turn the motor on and off. When the motor is turned on, current is going through the motor, and it starts to heat up. When you turn it off, the motor cools down. If you try to accelerate the duty cycle too much, you may inadvertently eliminate the cool down cycle, causing the motor to run without temperature cycling, therefore changing the end-use environment. Worse, you may shorten the turn on/off cycle so much that the motor actually starts to heat up and stabilize at a higher temperature than what it would normally experience in the field. If you do this, then you have to calculate not only the duty cycle but also the effects of increasing the temperature, not an easy calculation with motors.

RELIABILITY INTEGRATION: Integrating ALT with Your End of Life (EOL) Analysis

During the design phase, you predicted which components would reach its end of life during the useful life of your product. ALT is a great way to validate the EOL Analysis you performed to determine how close you were to your estimated values. If ALT shows that the EOL for a component is a shorter amount of time than you estimated in your EOL Analysis, you will either need to change your Preventive Maintenance (PM) strategy for that component, change the design, or change the component to increase the EOL. If ALT shows that the EOL for a component is a longer amount of time than you estimated in your EOL Analysis, you may be able to lengthen the time between PMs or even eliminate the PM for that component, depending on the ALT results.

CASE STUDY: ALT on a Router for a Police Car

A mobile router company needed to perform an ALT on the router they were developing for the police force to allow fast access to on-line information, as well as to allow all of the police cars to be able to communicate with one another. The end use environment for the router was in the trunk of a police car. Through analysis and research, we determined that the two portions of the product that would wear out first were the fans and the solder joints, and we determined that the dominant stresses to use to accelerate these would be temperature cycling, vibration, and humidity.

We set up two ALTs. The first was a fan ALT in which we used temperature and humidity as the primary stress and used the Peck's Model, a common humidity acceleration model. We increased the humidity from a 40% nominal/85% worst case to a steady 85%. The accelerant here was the humidity level. The second ALT was a solder joint ALT in which we used temperature cycling as the primary stress. For this, we kept the absolute levels the same as what we expected in the field, but we increased the duty cycle.

Note that we assumed that vibration would also be a factor, but vibration isn't often used during an ALT because of the difficulty in calculating the acceleration factor. Instead, we performed HALT on the entire product to show that the product had enough margins against the expected field environment. We achieved a significant amount of margin; therefore, we proved that vibration wasn't going to be the primary cause of failure, even though we couldn't accurately determine the actual life under vibration.

The results of the ALTs were that we proved the product would meet its expected field life.

37 HALT versus ALT—When to Use Each Technique

Determining when to use HALT and when to use ALT is the source of a great deal of confusion in the world of reliability. Now that we have introduced the topics of HALT and ALT, let's discuss when it's appropriate to use each technique.

HALT is a great reliability technique to use for finding dominant failure mechanisms related to design margins. However, in many cases, the dominant failure mechanism is wear-out. When this is the situation, you must be able to predict or characterize this wear-out mechanism to ensure that it occurs outside your customers' expectations and outside the warranty period. The best technique to use for this is a slower ALT. In many cases, you should use both HALT and ALT because each technique is good at finding different types of failure mechanisms. The proper use of both techniques together will offer a complete picture of the reliability of your product.

The Reliability "Bathtub" Curve in Figure 35.1 illustrates when HALT and ALT are most effective within the product life cycle. HALT is most effective in the steady state region, and ALT is most effective in the wear-out region.

37.1 Comparing HALT and ALT

Figure 37.1 shows a comparison between the two techniques.

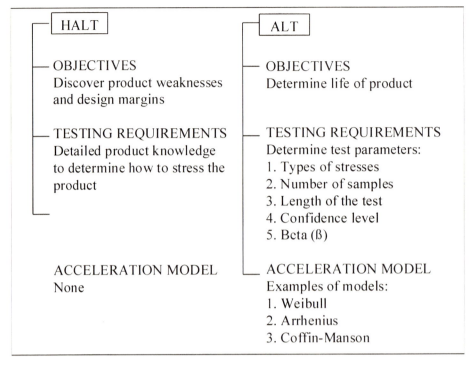

HALT	ALT
OBJECTIVES Discover product weaknesses and design margins	OBJECTIVES Determine life of product
TESTING REQUIREMENTS Detailed product knowledge to determine how to stress the product	TESTING REQUIREMENTS Determine test parameters: 1. Types of stresses 2. Number of samples 3. Length of the test 4. Confidence level 5. Beta (ß)
ACCELERATION MODEL None	ACCELERATION MODEL Examples of models: 1. Weibull 2. Arrhenius 3. Coffin-Manson

Figure 37.1: HALT versus ALT Comparison

There are three key areas in which we can compare HALT and ALT:

1. Objectives
2. Testing Requirements
3. Analytical Models

37.1.1 Objectives

The objective for HALT is to discover product weaknesses and design margins. As part of the process, you will experience failures and will need to perform a Root Cause Analysis (RCA) on each failure. You aren't as concerned about time to failure, or even the stresses that produced the failures, as you are about the failures themselves.

The objective for ALT is to determine the life of the product. The first step is for you to determine a Reliability Test Goal and then you need to identify the components with the dominant wear-out mechanisms so you can determine which components to test. In ALT, test the product at a constant level or using a cyclical stress. When you have failures, you are interested in the time to failure because this will be the basis of your life calculation.

37.1.2 Testing Requirements

For HALT, you will need detailed knowledge of the product in order to understand how to write your test plan and how to stress the product to find out the most information you can about the product.

For ALT, you need to determine specific parameters for the test in order to meet your goal. The parameters you will need to determine are types of stresses, number of samples, length of the test, confidence level, and Beta (β) (the slope of the Weibull Distribution). In our experience, you can model the wear-out of most components using the Weibull Distribution.

37.1.3 Acceleration Models

For HALT, you aren't interested in determining the length of the product life, and you won't be calculating acceleration factors, so there is no acceleration model.

For ALT, you will need to determine an appropriate acceleration model to be able to extrapolate the length of the product life properly.

37.2 Advantage of ALT over HALT

Three key advantages of ALT over HALT:

1. You Can Determine the Life of the Product
2. Samples Are Often Easier to Obtain
3. Test Equipment Is Often Less Expensive

37.2.1 You Can Determine the Life of the Product

Choose ALT over HALT when you need to know the length of life of the product. In HALT, you don't concern yourself with this much because you are more interested in making the product as reliable as you can, and measuring the amount of reliability isn't as important. However, with components that wear over time, it is very important to know the life of the component as accurately as possible.

37.2.2 Samples Are Often Easier to Obtain

Because we perform ALT at the component or assembly level, the samples are typically less expensive. In addition, they are usually available earlier in the program because you don't have to wait until you complete and integrate together different portions of the system.

37.2.3 Test Equipment Is Often Less Expensive

With ALT often you don't need any environmental equipment. An additional benefit for ALT is that you determine the product's life as well, where that's not typically the case for HALT.

We ran an ALT on a locking mechanism on a drawer for a medical cabinet, and we did so without using a HALT chamber. Instead, we used a very simple and inexpensive setup consisting of activating and de-activating the locking mechanism continuously. For this, we accelerated the duty cycle. With just that accelerant and a simple fixture, we were able to compress a ten-year test down to fewer than three days. See Figure 37.2 for a picture of the ALT fixture we built for accelerating a test on the locking mechanism.

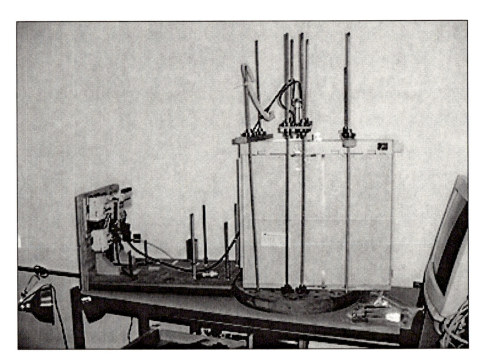

Figure 37.2: Fixture for ALT on Locking Mechanism

37.3 Advantage of HALT over ALT

Four key advantages of ALT over HALT:

1. Time Savings
2. Lower Number of Samples
3. No Acceleration Factor Calculation
4. Less Limitation on Stresses

37.3.1 Time Savings

With HALT, you can usually obtain the results in a shorter amount of time. You aren't as worried about time to failure as you are with which failure modes are dominant. You can usually find that out in a matter of days rather than weeks or months. This savings in time is also a big savings in money because it takes less time at a test lab or in a test chamber at your facility.

37.3.2 Lower Number of Samples

The number of samples required to perform a HALT is far fewer than with an ALT (usually just a few samples for HALT versus ten to twenty samples for ALT).

37.3.3 No Acceleration Factor Calculation

With HALT, you don't need to calculate an acceleration factor because the goal in HALT isn't to produce a reliability number, but rather to expand product margins.

37.3.4 Less Limitation on Stresses

With HALT, you aren't limited to using the same stresses as the field environment because in HALT you are only concerned with the failure, not how you produced the failure.

37.4 Combining ALT with HALT

Often you will run a product through HALT and then run the assemblies through ALT, where the dominant failure mechanism is wear-out.

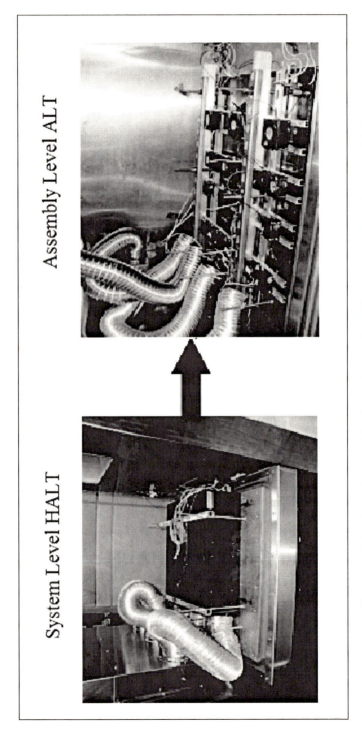

System Level HALT

Assembly Level ALT

Figure 37.3: Fan Assembly Run through ALT after System Level HALT

Chapter 37: HALT versus ALT—When to Use Each Technique

In Figure 37.3, we show a product in which we performed HALT at the system level. For most of the assemblies within the system, this was the appropriate level for finding issues. However, for the fan assembly, we didn't run the stresses long enough to excite the dominant failure mechanism, which was wear-out of the bearings. Therefore, we followed up the HALT with an ALT on the fan assembly by itself, and from the ALT, we were able to measure the life of the bearings.

At other times, you may develop the ALT based on the HALT limits, using the same stresses but lowering the stress levels so you can measure the amount of acceleration.

In Figure 37.4, we performed HALT on a power supply assembly, discovered the margins of the product, made several design changes as a result, then performed a Verification HALT to show improvements in the design margins. We then followed up the HALT process with an ALT on the power supply to measure the life of the product. Using this process, we discovered a key failure mechanism within the power supply that would have held up product release by months. We would not have found it had we performed only one of these two tests.

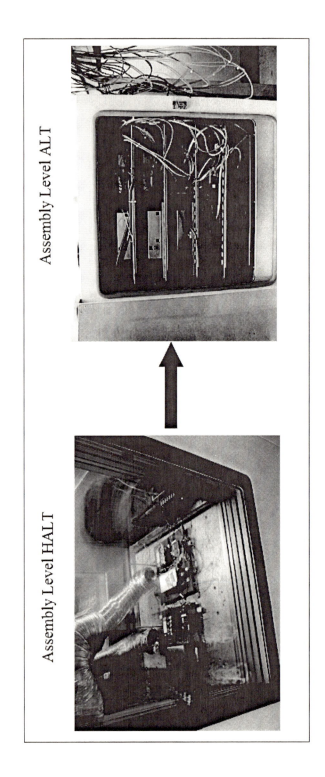

Assembly Level HALT

Assembly Level ALT

Figure 37.4: Power Supply Assembly Run through HALT and then ALT

RELIABILITY INTEGRATION: Integrating HALT and ALT Together
There are many times when it makes sense to use HALT and ALT together on a product. HALT is good at finding failure mechanisms related to design margins, whereas ALT is good at finding failure mechanisms related to wear-out, and most products will experience both types of failure types. During the FMEA, you need to determine which failure mechanism is going to be more prevalent, and if you determine that both types may occur, you need to set up tests to address each failure mechanism individually.

CASE STUDY: HALT and ALT on an Infusion Pump
An infusion pump company was developing a modified version of a product they had in the marketplace. This new project was to upgrade the original design with new features and to correct issues they discovered with the original design. Early in the program, we determined the differences from the previous design. We determined that the new design had the following additions:

1. Two new motors—one for the cassette insertion and removal, and one for the air sensor that pinches the tubing in order to detect air bubbles in the line.
2. A new power supply that was larger and also had a battery charging circuit for the new rechargeable battery.
3. A touch screen.
4. An 802.11 wireless feature.
5. A new rechargeable battery design.
6. New software to handle all of these new features.

We then performed an FMEA on each of these assemblies and determined which testing method was most appropriate for each new assembly. For items 1, 3, and 5, we decided that ALT was the best approach. For items 2 and 4, we decided that HALT was the best approach. For item 6, we decided that the best approach was to perform a Software FMEA followed by a rigorous review of the software requirements documentation, ending with the development of software test cases to exercise the requirements.

As a result of the testing program, we were able to launch the product successfully with a significantly lower failure rate than the original product release.

38 HALT-to-AFR Calculator

The **HALT-to-AFR Calculator** is a patent pending mathematical model that, when provided with the appropriate HALT and product information, will accurately estimate a product's actual failure rate (AFR) in the field. This model will also provide HASS or HASA data for the detection of an outgoing quality process shift in time from a stated AFR.

38.1 Background of Calculator

Since the concept of HALT was first invented, HALT practitioners have been looking for a way to calculate a field MTBF from the results of HALT. Many people have tried, but they haven't been able to produce meaningful results, either because their approach was incorrect or because there wasn't enough data to substantiate the model. There are three ways to approach the problem.

The first approach is using a Physics of Failure (POF) approach to model the failure behavior of each component in each environment. The problem with this approach is that there are far too many variables, and the POF model becomes too complex.

The second approach is using existing acceleration models and making general assumptions about acceleration factors or plotting best fit curves. The problem with this approach is that:

1. The acceleration factors are usually not correct and can't be generalized.
2. Current acceleration models are set up for constant stressing and not step stressing. With step stressing, you don't dwell long enough at each stress to be able to accurately extrapolate the reliability.

3. There are typically not enough failures that occur to make the data statistically significant, so when you attempt to use current acceleration models, the confidence intervals (the upper and lower values of the confidence levels) become unrealistically wide, and the conclusions of the data become erroneous. We once tried using one of the more popular Weibull Models on HALT results and achieved the result of an MTBF of 100,000 hours with a lower confidence of 1,000 hours and an upper confidence of 10,000,000 hours. Because the confidence intervals were so wide, the data was useless.

The third approach is to collect a lot of data on products that have gone through HALT and their subsequent failure history in the field, then build a mathematical model to correlate the HALT results with the field results. The problem with this approach is that it requires a lot of data from many different types of products in many different industries to develop an accurate model. In the past, no one had enough data to build a model.

Fortunately, the developers of the HALT-to-AFR Calculator were able to collect enough data to develop an accurate model. They collected data from HALT and subsequent field results for the past fifteen years that include over fifty different products from twenty different industries. The developers continue to collect data and add to the model, thereby making it more accurate over time.

38.2 Benefits of the Calculator

The main benefits of the HALT-to-AFR calculator are:

1. Performing HALT followed by running the calculator takes significantly less time and money to run than a Reliability Demonstration Test (RDT). The application of this calculator can be a huge time and cost saver.
2. The calculator provides recommended minimum stress levels for HALT. These levels can give you assurance that your product will exceed your customers' expectations and allow you to forecast warranty expenditures accurately.
3. The model can accommodate HALT samples sizes from one to six, with the optimum size being four. This quantity of systems means that there should be four systems in each environment.
4. The model can calculate ninety percent upper and lower confidence limits.

The calculator will estimate HASS parameters for production. The inputs for HASS and HASA are the daily sample size and the detectable shift in the AFR you wish to detect.

38.3 How to Increase the Accuracy of Your HALT-to-AFR Calculation

Here are a few ways you can increase the accuracy of your HALT-to-AFR calculation:

1. Perform Root Cause Analysis (RCA) and provide corrective actions for all issues encountered within the Guard Band Limits. The Guard Band Limits are the limits that we strongly recommend you meet as a minimum. From our experience, products that can't meet the Guard Band Limits have significantly lower reliability than products that can. Note that we still recommend you perform HALT to the Fundamental Limit of the Technology (FLT) whenever possible.

2. Prior to starting the HALT, remove or bypass any circuitry or firmware being used to protect the product from over temperature or excessive vibration. For example, many hard disk drives have accelerometers mounted on them to detect vibration which then triggers the protection mechanism to park the heads of the drive. If the circuitry or firmware is protecting an assembly or component of the product that will be destroyed when going beyond the cutout stress point, then locate the assembly or component external to the chamber or suspended away from the vibration table or the thermal airflow. By doing this, you can continue the HALT to verify the true robustness of the product. Otherwise, the AFR estimate may be artificially high because the calculator will estimate a higher failure rate than what will really occur in the field.

3. Prior to starting the HALT, build extender cables to separate the stress sensitive assemblies (for example, an LCD display) and mount them outside the chamber. If you test to the FLT and still do not meet the Guard Band Limits, use the extender cables to separate out these assemblies and continue testing. If you don't build extender cables, the testing will be limited by these early failures. If this happens, you won't discover other potential failure mechanisms, and the AFR estimate may be artificially high. If you are able to build an extender cable and separate out an assembly that failed due to reaching its FLT, you can dismiss this failure from the calculation. By definition, when you reach an FLT, you haven't accelerated a failure that will occur in the field over a longer period of time. Rather, you have caused a new nonrelevant type of failure. If it is not possible to build extender cables, consider changing the technology to one less sensitive to stress.

4. Perform HALT with a protocol that sufficiently tests the product in each stress environment. We recommend seventy-five percent test coverage as a minimum. Otherwise, the AFR estimate may be artificially low because customers will detect failures modes that you missed during HALT.

If you don't have adequate test coverage, then we recommend you perform HALT again later in the product life cycle when you have better test coverage and can use these newer results for your calculation.

38.4 Limitations of the Calculator

The calculator has many benefits, but there are a few limitations:

1. The model can't estimate wear-out mechanisms. You will need to address these types of failures using Accelerated Life Testing (ALT).

2. At this time, the calculator can only take into account the stresses of temperature and vibration. We discuss in Chapter 34 how you shouldn't limit HALT to just these stresses, but rather you should use any stresses you deem appropriate for finding product weaknesses. We still recommend you do this, but we can't include this extra information into the model at this time. In the future, you may be able to add other stresses to the model if you can collect enough data on these additional stresses.

3. You need to perform HALT using a protocol that sufficiently tests the product in each stress environment. If the test coverage can't find an issue, it can't be included in the model.

4. The calculator is new and does not have data from all types of products. If your product type is significantly different than the types of products used to build the model, your results may not be as accurate as products that are more similar to those used to build the model.

38.5 How to Use the Calculator

In order to use the HALT-to-AFR Calculator, you will need to fill in the following information shown in Table 38.1.

Table 38.1: AFR Calculator Input Table

	Input Matrix	
MTBF (in Hrs) =	A	User Input
Product Thermal (Hot in °C) =	B	User Input
Product Thermal (Cold in °C) =	C	User Input
Product Vibration (in Grms) =	D	User Input
Prod Published Spec Level (see below) =	E	User Input
Number of HALT Samples =	F	User Input
HASS or HASA (yes = 1, no = 0) =	G	User Input
If HASS or HASA, Daily Sample Size =	H	User Input
If HASS or HASA, Detectable Shift in AFR (in %) =	I	User Input
Steady State AFR, % (HALT Only) =	AFR	Calculated Field
Steady State Field MTBF, Hrs (HALT Only) =	MTBF	Calculated Field
Lower 90% HALT Confidence Limit =	MTBFlower	Calculated Field
Upper 90% HALT Confidence Limit =	MTBFupper	Calculated Field
Days to Detect Shift w/ HALT/HASS/HASA (Max) =	Shift Value	Calculated Field

A. Enter the MTBF estimate (in hours). You can use the methodology of a Reliability Prediction described in this book to calculate the MTBF. You can use any industry accepted technique for this calculation, including Telcordia or a similar tool. If this estimate isn't available, use 40,000 as a default value. This parameter has only a small effect on the final field AFR estimate due to the highly variable processes followed by the many assumptions used in estimating an MTBF.

B. Enter the final hot operating limit (HOL) (in °C) achieved in HALT as measured on the product and not the chamber set point.

C. Enter the final cold operating limit (COL) (in °C) achieved in HALT as measured on the product and not the chamber set point.

D. Enter the final vibration operating limit (VOL) (in Grms) achieved in HALT as measured on the product and not the chamber set point.

E. Enter the product's published thermal operating specifications (in °C). Try to match your product's published specifications to a corresponding level number listed in Table 38.2. For example, a high-end consumer product equates to a Level 2. Note that the category column description may not exactly match your product's field application, but choose the column that most closely fits.

Table 38.2: Product Published Specification Level

Published Spec, °C	Level	Application	Guard Band, °C
0 to +40	1	Consumer	-30 to +80
0 to +50	2	High-end Consumer	-30 to +100
-10 to +50	3	Hi Performance	-40 to +110
-20 to +50	4	Critical Application	-50 to +110
-25 to +65	5	Shelter	-50 to +110
-40 to +85	6	All Outdoor	-65 to +110

F. Enter the number of systems used in the final HALT only. This isn't the total quantity used in all of the HALT efforts.

G. Enter 1 if you will be performing HASS or HASA on the product.

H. If 1 is selected for entry G, enter the daily sample size. The daily sample size is the number of systems that you will subject to the HASS or HASA process in a twenty-four hour shift. If the HASS or HASA process control chart varies dramatically from shift to shift, then use an eight-hour shift sample size until the control variables are under statistical control.

I. If 1 is selected for entry G, enter the detectable shift rate as a percentage. The detectable shift rate in AFR is the delta between the outgoing AFR and the detectable shift in outgoing quality (from HASS or HASA) that you wish to detect. For example, if the product baseline AFR is 4% and the worst case AFR is 10%, the detectable shift rate would be 6% and you would enter a 6 in this cell.

Once you complete the data input for the model, the calculator will automatically calculate the AFR, MTBF, confidence limits, and days to detect shift in AFR if you are using HASS or HASA.

Table 38.3 shows an example of a HALT-to-AFR Calculator table filled out.

Table 38.3: HALT-to-AFR Calculator Example

	Input Matrix
MTBF (in Hrs) =	40,000
Product Thermal (Hot in °C) =	111
Product Thermal (Cold in °C) =	-80
Product Vibration (in Grms) =	55
Prod Published Spec Level (see below) =	3
Number of HALT Samples =	4
HASS or HASA (yes = 1, no = 0) =	1
If HASS or HASA, Daily Sample Size =	64
If HASS or HASA, Detectable Shift in AFR (in %) =	2
Steady State AFR, % (HALT Only) =	0.46
Steady State Field MTBF, Hrs (HALT Only) =	1,886,724
Lower 90% HALT Confidence Limit =	1,017,137
Upper 90% HALT Confidence Limit =	3,878,012
Days to Detect Shift w/ HALT/HASS/HASA (Max) =	59

The developers of the calculator are constantly adding data to the model to make it more accurate. If you have HALT data along with corresponding field data that you can supply for the developers to add to the model, please contact me and I can put you in touch with the developers.

RELIABILITY INTEGRATION: Integrating Reliability Prediction and HALT with the HALT-to-AFR Calculator

Two key pieces of information we use for the HALT-to-AFR Calculator:

1. MTBF results from the Reliability Prediction
2. Operating limits from the HALT

The HALT-to-AFR Calculator uses the MTBF number from the Reliability Prediction. We recommend that you compare the highest failure items in the Reliability Prediction to the failures you found in HALT. The better the correlation between these two, the more accurate the HALT-to-AFR Calculator will be.

The HALT-to-AFR Calculator uses the operating limits from the HALT, which you directly enter as a data point. The more thorough you are in HALT in expanding the design margins, separating out weaker assemblies, removing protection circuitry or firmware, and providing effective corrective actions, the more accurate the HALT-to-AFR Calculator will be.

CASE STUDY: Switching from RDT to HALT-to-AFR Calculator

Many of our clients have commented to us that they can't justify using RDT because RDT requires too many samples and the testing takes too long to get the results. Yet HALT in itself isn't capable of determining field MTBF. This is precisely why the HALT-to-AFR Calculator was developed. We have deployed it with over a dozen companies converting them from RDT to the HALT-to-AFR Calculator, and the results have been very positive.

39 The Difference between Fixing Failures and Taking Failures to Root Cause

Root Cause Analysis (RCA) is the investigative process to determine the underlying event(s) responsible for a failure. Failures are associated with component integrity, proper functioning of a complete system, or the execution of an engineering process. They are most often classified as being either mechanical, electrical, or software in nature. There are other possible classifications such as chemical, structural, and optical, but in this chapter, we will concentrate on mechanical, electrical, and software RCAs. The underlying "root cause" event can be associated with the design, manufacturing, or end usage conditions, as well as other elements of a system's design.

Failures can prevent you from achieving your reliability goals. Nevertheless, if we can learn from these failures, we can actually make the product more robust. These failures can be of a unique nature, requiring specialized techniques and methodology to properly diagnose and understand the root cause. Your organization's internal resources may be limited in capability and not properly equipped to address all failures your products may be experiencing. That is when it is appropriate to turn to outside resources.

RCA is the foundation for every effective reliability program. You can leverage RCA in both the development of new products and the reliability improvement programs associated with an existing product base. The FMEA and FTA reliability analysis techniques, as well as the ALT and HALT reliability testing techniques, are all designed to identify (potential) failures. RCA is the essential follow-up activity that must be effective in order to resolve and eliminate the failures. Similarly, the Closed Loop Corrective Action (CLCA) process is a database process that identifies and prioritizes the failures going through the RCA process. Effective RCA programs not only improve reliability performance but also reduce warranty costs that improve your organization's profitability.

39.1 Seven Step Process

We recommend the classic 7-Step RCA Process:

1. Define/identify the problem.
2. Analyze and gather data/evidence.
3. Determine root cause. Ask why it occurred, and identify the causal relationships associated with the defined problem. Identify which causes, if removed or changed, will prevent recurrence.
4. Choose solutions and an action plan. Identify effective solutions that prevent recurrence, are within your control, meet your goals and objectives, and don't cause other problems. The solutions shouldn't only address the problem at hand, but should also ensure that similar problems don't occur on this product or on other products prone to the same type of problem.
5. Implement solutions.
6. Evaluate. Observe the recommended solutions after they have been implemented to ensure they were effective.
7. Report on what you learned.

Note that some organizations have an eighth step to the process and call this the eight-discipline (or 8D) process. The eighth step is the step where the organization celebrates for having completed the RCA. I don't agree with this method of celebrating because you may be rewarding the same team that allowed the error in the first place. A better reason for celebration is if there are no RCA events in a specified period of time.

As part of the documentation of the issue and entry into the CLCA process, you should also determine the priority of the issue and the relevancy of the issue. These will play a part in how you approach the failure analysis.

There are many different products and technologies that undergo RCA. Here I will cover three of the main categories:

1. Electrical RCA
2. Mechanical RCA
3. Software RCA

39.2 Electrical RCA

The following are examples of situations requiring Electrical RCA:

1. Printed circuit boards (PCBs)
   a. Identify manufacturing defects
      i. Conductive anodic filament (CAF)
      ii. Plated through hole (PTH) fatigue
      iii. Electromigration
   b. Identify handling issues
      i. Electrostatic discharge (ESD) effects
      ii. Humidity failures
      iii. Shipping damage
2. Interconnects
   a. Solderability issues
   b. Intermetallic formation
   c. Wear-out failures due to stresses such as thermal cycling or vibration
3. Die-level issues
   a. Passivation cracking
   b. Die cracking
   c. Electrical overstress (EOS)
   d. Electromigration
   e. Dielectric breakdown
   f. Hot carrier injection

During the process, there are many different non-destructive and destructive failure analysis tools and techniques you can utilize:

Non-Destructive Techniques

1. Optical microscopy
2. Electron microscopy
3. Ion chromatography
4. Surface analysis techniques such as:
   a. Fourier transform infrared (FTIR) spectroscopy
   b. X-Ray fluorescence (XRF)

5. Material analysis techniques such as:
    a. Thermo-mechanical analysis (TMA)
    b. Thermo-gravimetric analysis (TGA)

Destructive Techniques

1. Cross-sectioning
2. Decapsulation

Always start with non-destructive techniques. Move to destructive techniques if you can't conclusively determine the cause using non-destructive techniques. Because non-destructive techniques don't destroy the evidence, you can apply a series of non-destructive techniques in search of the root cause. However, once you move to destructive analysis, you typically only have one chance of determining the root cause before you destroy the evidence.

39.3 Mechanical RCA

The following are examples of situations requiring Mechanical RCA:

1. Mechanical wear-out of joints, lubricants, and bearings
2. Structural fatigue

Many of the failure analysis techniques listed for electrical-related RCA are also used in RCA of mechanical components. Here are a few techniques unique to mechanical components:

Non-Destructive Techniques

1. Stress analysis
2. Fatigue and fracture mechanics analysis
3. Creep degradation analysis
4. Nonlinear finite element analysis (FEA)
5. Computational fluid dynamics (CFD)
6. Probabilistic evaluations
7. Micro-testing

Destructive Techniques

1. Bend testing
2. Pull testing

39.4 Software RCA

RCA with software has both similarities and unique differences compared with hardware. Software RCA is different in that finding a software bug and fixing it (analogous to finding a hardware failure and redesigning it) is typically resolved within the software group, and rarely do you involve departments outside of software or resources outside of your company. With Hardware RCA, often you may need to pull in other departments to help resolve the issue, and occasionally you may need to pull in resources outside of your company, such as a failure analysis lab. Where Software RCA is similar is that once you fix a bug, you need to implement new techniques, such as the ones we discussed in Chapter 31, including Facilitation of Code Reviews, Software FMEAs, Software FTAs, and Software Phase Containment Metric Tracking to prevent this particular class of bug from reappearing. This is similar to the corrective action portion of the Hardware RCA, where you not only need to fix the problem and prevent that particular problem from recurring, but you also need to fix the process that caused the problem in order to show continual improvement. See Chapter 31, Software Reliability, for examples of best practices you can use to improve your Software Failure Analysis process.

RELIABILITY INTEGRATION: Integrating Design Techniques during RCA

Some of the best reliability techniques used during the design phase can also be used during RCA. Examples of these are Design of Experiments (DOE) and Fault Tree Analysis (FTA).

In the design phase, DOE is a great technique for determining the optimal combination of factors from a list of possible factors. During an RCA, if you can't determine how to fix the problem, a DOE may be helpful in coming up with the optimal solution.

In the design phase, FTA is a great technique during Reliability Apportionment to allocate reliability to assemblies properly. During an RCA, you can use FTA by starting with the end effect then drawing an FTA tree. The first branch of the tree would be the possible major assembly or functions that could have caused the end effect. The next branch would be the possible failure modes for each major assembly. As you draw the tree, you can start eliminating branches of the tree if they don't match up with the RCA evidence. You can then narrow down to a smaller subset of possible failure modes that could have caused the end effect.

CASE STUDY: Mechanical RCA on a Solar Structure

Our client had a fracture that developed on a mechanical portion of a solar panel structure and needed to determine if it was related to design, material, or manufacturing process, or potentially due from the environment (high winds could have possibly added to the stress).

Client Attempts to Fix Problem: The client reviewed wind reports and various other reports and suspected that either the wind analysis was performed incorrectly or that inferior steel was used during manufacturing.

The Root Cause Analysis: We visited the site, examined all of the evidence at the site, took lots of photos, and took back failed pieces. We put these samples through material analysis, and we calculated the loads and reviewed the wind report.

Techniques Used: Material Analysis, Finite Element Analysis (FEA).

Findings: The observed damage pattern in conjunction with FEA and structural testing indicated a very high probability of damage due to over-tightening of portions of the structure during installation. We also discovered a design change by our client that weakened the structure; this change had been made after they performed the original wind analysis. During inspection, we found that the installers omitted the lock washers on the main bracket adjusting bolts, even though they were called out on the installation drawing. In addition, we determined that there was a flaw in the algorithm for adjusting the tracker, because it was programmed to adjust until the tracker reached the requested position without checking how far it traveled. When the structure bent, the motor could no longer move the tracker, but the motor kept driving the tracker and ultimately caused much more damage than what originally would have taken place.

Resolution: Our client performed the wind analysis on the new design, modified the tracking algorithm, and modified the training procedure for installing the system.

40 The Corrective Action System—The Backbone of a Good RCA

The corrective action process has different names in different industries. Some call it Corrective and Preventative Action (CAPA), some call it a Failure Reporting Analysis and Corrective Action System (FRACAS), and others call it **Closed Loop Corrective Action (CLCA)**. In this book, we have standardized on the acronym CLCA. Whatever you call it, it is essentially the same process—to identify, analyze, and correct a problem with a product or process using a systematic approach to prioritize each problem so that each are dealt with on a timely basis. During all of the testing, you should utilize a robust CLCA process to capture, analyze, and, if necessary, correct all failures you discover.

The objective of a CLCA is to provide corrective action in a timely manner for any problem, and to have a means of verifying that the failure did indeed fix the problem. A good CLCA process is vital to your organization. It is the method by which you track and fix problems in a methodical fashion.

The steps of a CLCA process:

1. Plan/Outline of Process
2. Choice of Failure Analysis Software Tool
3. Implementation of Failure Analysis Software Tool
4. Education/Training

40.1 Plan/Outline of Process

Develop a plan that outlines a failure analysis process tailored to your company and situation. This plan will identify the flow of the process from each area in which a failure can occur, including (but not limited to) design, prototype test, manufacturing, and customer site. This plan will also identify all of the resources necessary to make the process successful on an ongoing basis, including identifying responsible parties for signing off the corrective action.

40.2 Choice of Failure Analysis Software Tool

Once you have identified the process, it is time to choose a software tool to use for the process. You can use anything from a simple spreadsheet program to a more complex relational database. Special modules are also available from various reliability software tool vendors. You should choose the best tool for your particular application. If none fit your needs well enough, you may need to develop your own.

40.3 Implementation of Failure Analysis Software Tool

If you choose a tool from a software tool vendor, you will need to work with your vendor to implement it within your organization. Make sure the appropriate "hooks" are put in place so that it integrates with your other software tools, such as your customer relationship management (CRM) database and your product lifecycle management (PLM) system. This tool should also be linked into your Field Data Tracking System.

40.4 Education/Training

You will need to educate key individuals on your staff on how to use the tool and how to manage the process effectively. If you purchased a product from a software tool vendor, they usually have training classes on their tools.

You will also need to educate your field support team, your key suppliers, and your customer. You should make sure to train them on the proper procedure of tagging and returning items, especially on what information is critically needed for successful Root Cause Analysis (RCA) on specific items and for data trend analysis in general. This information includes date, time to failure, serial number of failed assembly, serial number of system, failure symptom, and failure resolution.

RELIABILITY INTEGRATION: Integrating CLCA with RCA
It is important to integrate your CLCA process with your RCA process. If you have a good RCA process but it isn't integrated with your CLCA process, you will repeat the same types of failures over again because they won't be well documented and communicated well to others. By integrating the two processes together, you will create a powerful method for correcting problems and ensuring they don't recur.

CASE STUDY: Instituting a Robust CLCA System
A networking company had a disjointed corrective action system and always seemed to be in the reactive mode. We helped develop a plan that outlined a CLCA process fit for them, identifying each area in which a failure can occur as well as identifying all of the resources necessary to make the process successful on an ongoing basis. We then helped them develop and implement a simple but customized failure analysis tool that tracked each failure through the entire process; we also created a field failure tracking tool that linked to it. We then trained the appropriate personnel on how to use this new tool. The result was that this new process sped up the whole corrective action process, which freed up other resources to concentrate on new designs.

Part V
Manufacturing Phase

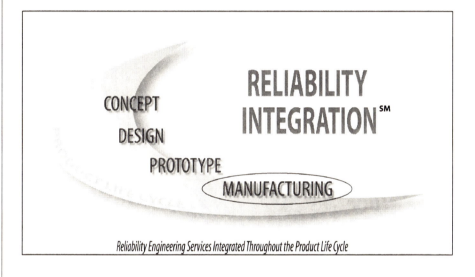

RELIABILITY INTEGRATION℠

CONCEPT
DESIGN
PROTOTYPE
MANUFACTURING

Reliability Engineering Services Integrated Throughout the Product Life Cycle

41

Assess Your Vendors as You Assess Your Own Organization

A **Vendor Assessment** is a systematic evaluation of a broad range of potential reliability activities and techniques as currently employed and integrated with one or more vendors.

When your product's reliability is critically linked to your vendor's reliability performance, your team will need to have knowledge of your vendor's current reliability program and how best to make changes to improve their product reliability. A few reasons you may choose to assess one of your vendors:

1. Increasing field returns of the vendor's product
2. Changes in your customer expectations
3. Your company wants its products to compete on reliability
4. Need to reduce warranty costs
5. Bringing on a new vendor

Of the many potential reliability activities to improve or introduce, which ones will fit with your vendor's organization and provide the desired return on investment?

The methodology of the Vendor Assessment is similar to the Reliability Program Assessment described in Chapter 5, except that you will be assessing an outside organization rather than your own. The main difference will be in the implementation of the action items. Within your own organization, you have much more control over what type of corrective action to take and the speed at which to implement.

A Vendor Assessment provides an objective view of your vendor's existing reliability program and permits the effective investment in areas of their program that will efficiently improve their product reliability. From an understanding of your vendor's current reliability program, develop a set of short and long term actions that will significantly improve your vendor's ability to develop and produce reliable products. You need to focus rapidly on improvement efforts of the critical few items coupled with a long term view and plan to get there. This will assist your vendor to alter their reliability program's capability dramatically.

RELIABILITY INTEGRATION: Integrating Competitive Analysis with a Vendor Assessment

If you find your vendor's performance is below where you would like it to be when assessing a vendor, work with your vendor to help them improve those identified areas. What if you can't get their level of reliability where it needs to be for your program? It may be time to look for an alternate vendor. Once you identify a second vendor that meets your performance, cost, and schedule criteria, it is time to perform a Competitive Analysis between the two vendors. See Section 7.1.3 for more information on Competitive Analysis. If the new vendor outperforms your current vendor, you have two options:

1. Show the results to your current vendor to get them to pay more attention to reliability and to start making improvements to their product or process.
2. Switch vendors.

If the new vendor doesn't outperform your current vendor, you need to continue to look for alternate vendors. To save time, you can perform a Competitive Analysis on several vendors at the same time and choose the one that scores the highest.

CASE STUDY: Assess Your Vendors to Improve Manufacturing Yields
A telecom company was experiencing a high failure rate both in production and in the field. After we performed a Reliability Program Assessment and analyzed their field return data, we found that a significant source of defects came from the printed circuit assembly (PCA) provided by its vendor. The current process of returning boards to their vendor for Root Cause Analysis (RCA) hadn't delivered improvements. We then followed up with a Vendor Assessment on their contract manufacturer (CM). We focused the assessment on proactive and systematic changes to the manufacturing process. The results were that our client and their CM found areas of improvement in communication, board layout, and manufacturing processes. By working together, we helped them cut their workmanship related field issues in half within six months. Our client's vendor also benefited by having a better product for its other customers.

Chapter

42 Outsourcing versus In-House Designing and Manufacturing

Today, more and more companies are outsourcing their design and manufacturing processes. When you outsource your design and manufacturing, you are outsourcing your reliability.

Nevertheless, this doesn't mean that you have to lose your ability to control your quality. It just means you have to use new and innovative methods to ensure good quality. Effective communication is key. Face-to-face contact is essential. Here are five main risks you may face when you outsource:

1. Lack of Control over the Design Process
2. Lack of Control over the Manufacturing Process
3. Lack of Visibility into the Testing Process
4. Lack of Visibility in Correlating Field Failures to Production Failures
5. Lack of Visibility When Changes Are Made

For each of these five areas, I will describe the risk as well as discuss how to mitigate the risks.

42.1 Lack of Control over the Design Process

When you outsource your design, you are using an original design manufacturer (ODM). You typically outsource to your ODM the areas of your design that are more common in the industry, that is, areas not related to your core intellectual property (IP). Because your ODM is responsible for portions of the design, your ODM will also control the reliability of those portions of the design. This means that you should include your ODM's portion of the design during your Reliability Apportionment and provide the apportioned amount to your ODM in the form of a reliability goal. Your ODM should create a Reliability Program Plan (RPP), then they should execute this plan. However, they may be unfamiliar with this process; therefore you will need to educate them on how to create and execute one. Even if they are familiar with the process, you will need to work with them to ensure their plan is synchronized with the plan for your overall product.

42.2 Lack of Control over the Manufacturing Process

When you outsource your manufacturing, you are using a contract manufacturer (CM), also referred to as an electronic manufacturing service (EMS). Every production line has an inherent failure rate, often referred to as the process capability (CPK). It is important to understand exactly what is contributing to the yield loss. Pick and place equipment (equipment used to install components onto a circuit board), test equipment, and even the operators affect the yield. Because you don't own the process or equipment, it is very difficult to characterize it and identify which factors are contributing to the yield loss.

Failures can be generated by the production equipment and/or personnel. For example, electrical overstress (EOS) caused by electrostatic discharge (ESD) is created by many sources, such as poorly grounded equipment or people not properly wearing grounding straps.

Inventory control is a problem with outsourcing. Obtaining information on inventory levels is difficult, especially if you need to determine inventory levels by date codes. Date codes are the product build dates that vendors stamp onto their product. These date codes can be very useful if you discover a failure on the manufacturing line or in the field that may be related to a particular date range of when the product was created. When you or your CM identify material problems, you may have difficulty quarantining the suspect parts while you are conducting the investigation. Often, your CM must return the suspect parts or boards to your engineering group for the evaluation. Because it can take days to get the faulty material in to the hands of your engineer, you lose valuable time. For critical failures, you may need to hand-carry the samples back to your facility.

42.3 Lack of Visibility into the Testing Process

Often the data from the manufacturing test process can be spread across different databases, and reports may be lacking in complete and real information. You may not notice that data is duplicated, which results in a lot of data but not much useful information. For these reasons, you may have difficulty analyzing the data to provide effective corrective action. In addition, you may have difficulty determining if any previous corrective actions that your CM implemented were effective.

First pass yield information (the percentage of samples that pass the testing process the first time through without requiring any rework) is an important indicator of the health of a test process. However, by the time you obtain first pass yield data, you probably have already shipped the material to customers.

How do you know the number of times a sample has been tested and repaired before it passes the test? Most companies have a rule on how many times they allow this. If you don't have visibility to the test process, you won't be able to verify this. In addition, how do you assess the quality of the rework? All of these concerns can present difficult decisions when outsourcing. No problem found (NPF) or false failures can be difficult issues to resolve. You may receive reports of intermittent failures being shipped to customers because the customers returned the assemblies a second time or, in some cases, a third time, and each time they tested NPF. At times like these, we recommend implementing HASS in the repair center to find these intermittent failures (see Section 43.3).

Without the proper test verification and failure analysis, the assemblies that end up in the bone pile (the pile of assemblies that can't be fixed readily, so they are set aside to be looked at sometime later in the future) can build up quickly. Most CMs get paid for the assemblies they ship and this is where they put most of their efforts.

In addition, setting up a manufacturing screening process can be tricky, and if it is not done properly, you can actually be inducing failures. Good communication is required between you and your CM for this to be implemented successfully. You should monitor the manufacturing screening process because the failures from this process will be a sign that your CM's process is shifting. Often, you will want to tweak the manufacturing screening profile to optimize it, especially at the beginning of the manufacturing process. After you change the design, you may need to change your manufacturing screen and you will usually have to reprove the screen. You will likely need to be at your CM's site to guide them through this process. See Chapter 43 for an explanation on HASS, manufacturing screening, and how to reprove a screen.

42.4 Lack of Visibility in Correlating Field Failures to Production Failures

One great method of continual process improvement is to monitor the field results dramatically and then to make changes accordingly. This may require changing the testing process if you determine the cause of an escape (a defective product that passes the test) was due to lack of test coverage. Alternatively, the mistake may have been something that testing can't find but inspection is the answer. Correlation of field failures to the test data is an important part of process improvements. You need to determine if production testing is screening out the failures, or if the production testing is generating the failures. Having the ability to run the material that failed through the production testing again to see if it fails the second time through is extremely important, but it is difficult to manage with a CM. If the field failure passes production testing the second time through, then this would indicate that your testing of the product is not catching the errors. You then need to look at improving your test program. Are the proper people in the proper place to manage this? You will need to be at your CM frequently to help manage this process.

42.5 Lack of Visibility When "Improvements" Are Made

Improvements come in various ways. In some cases, it is personnel-related; in others, it is equipment-related. Improvement can come from changes in process flow and the handling of the material, as well as how your CM stores and inventories material. Your CM is supposed to notify you of all changes they make, but they may not do this, or what they tell you they are doing may be different from what they are actually doing. In addition, it is important to correlate the effects of changes on the manufacturing floor. However, you may have difficulty doing this with your CM because there will be a delay in obtaining the data necessary to properly assess the effects of change. When you get the data, the data may be incomplete. You need to be on their manufacturing floor frequently to ensure that they carry out changes and to ensure they don't make other changes without your knowledge.

42.6 Summary

Managing your outsourcing means you must take an active role with your ODM and your CM, and you should visit them often to ensure processes are running smoothly and correctly. It may be more expensive at first, but the cost will go down once you establish the process, and it will ultimately save you money in reduced production and field failures.

RELIABILITY INTEGRATION: Integrating Your Reliability System with Your Vendor's System

One common mistake companies make when they outsource their design or manufacturing process is that they don't follow up with their vendors and assess their vendors' reliability and quality programs. Some companies choose to take the role of police officers, inspecting and testing the output of their vendors. Instead, we advise you to work with your vendors early on and help them develop a reliability program that is synchronized with your own. You should be involved in your vendors' review meetings having to do with your product. Keep track of their Reliability Metrics so that there are no surprises during testing or, worse, after you start shipping the product.

CASE STUDY: "Tell Us What to Do"

We were working with a startup company that was outsourcing the majority of their product design to an ODM. In discussions with the ODM, I asked them to show me what they intended to use for a Reliability Program Plan (RPP) and their response was, "We will do whatever you want us to do." I responded by saying "I don't want to tell you what to do; rather, I would like for you to tell me what you think should be done and why." Again they told me they would be more than happy to follow my instructions. After a few rounds of this I finally got them to show me their proposed plan, which as it turned out was copied straight from a plan from one of their multi-billion dollar customers. Clearly we weren't interested in using this type of plan—we didn't have the time or the money to follow such a comprehensive program, plus we were producing a product for a completely different market than their customer's plan was intended for. I sat down with their team and explained to them that we weren't looking to tell them what to do every step of the way. Instead, we wanted them to decide what the best program was and they could utilize us for guidance as needed. As long as they had a good rationale for their reliability program, we would go along with it. They finally realized what we wanted from them, and they developed a cohesive plan. We accepted the plan with only minor revisions. The ODM then executed the plan, and we guided them as needed. This process resulted in one of the most reliable products our client ever produced. As a result of this interaction, the communication between my client and the ODM is much better, and the expectations from both sides are much clearer.

43 All Screens Weren't Created Equal—The Importance of HASS

Highly Accelerated Stress Screening (HASS) is a process comprising of a set of stresses performed on a product before it is shipped, with the goal of finding manufacturing related defects. The set of stresses combined together make up the screen. What makes HASS different from most other types of manufacturing screens is that, with HASS, you choose the optimal screen for the product.

An optimal screen means that the screen levels are high enough so as to find defects but not so high as to start causing other parts of the product to wear out. This process of optimizing the screen is called Proof of Screen (POS). We will cover how to optimize the screen later in this chapter.

Note that we are using the term HASS to imply using any type of stress that can accelerate finding potential manufacturing defects. As with HALT, many people equate HASS with using temperature and vibration. In fact, you can use a variety of stresses during HASS, but only *if* you can prove that the stress is capable of finding a defect in an accelerated manner.

In today's world of electronic and electromechanical systems, the reliability of components is becoming continuously better; therefore, some of the old methods of screening that were effective at finding component defects aren't as effective as they once were. What engineers need is a more sophisticated method of screening whereby the screen is tailored and tuned to process weaknesses. This is what HASS is all about.

Most companies have some form of environmental stress screening that they are already using, such as the following:

1. Run-in—powering on a system and running tests without the use of accelerated stresses. The key variable is the run-in time.
2. Burn-in—running a system at elevated temperature to take advantage

of heat as an accelerant. The key variables are the burn-in temperature and the amount of burn-in time.

3. Temperature cycling—cycling the temperature of a system between cold and hot. The key variables here temperature ranges, temperature rate of change, dwell time at each temperature extreme, and the number of cycles.

4. Vibration—applying a vibration stimulus to the system. The key variables are the vibration level, whether the level is constant, the frequency range of the vibration, the amount of energy at each frequency level, and the amount of time.

5. Power cycling—cycling the system power on and off. The key variables are the time between cycles and the number of cycles.

You shouldn't confuse these types of stress screens with HASS because HASS isn't a type of stress, but rather a methodology. In fact, you may choose one or more of the screening methods as part of your HASS methodology, but in order for it to be HASS, you must prove that you have optimized the screening methods you have chosen.

One variable that is common to all of these is the amount of test coverage during the screening. This is probably the most important variable and often is overlooked. The higher the test coverage, the better. See Section 33.2 for a discussion on test coverage.

43.1 HASS versus Burn-In

Many companies still use burn-in as a screening methodology. They do so not because they have proven that the screen is optimal, but rather "because that is the way it has always been done." There are certain types of defects that burn-in is capable of finding—new classes of components that have immature processes—and for these, it is possible to prove that burn-in is the correct type of screen and can be incorporated into a HASS methodology (in order to prove it, you must optimize the burn-in profile). However, for most other types of defects that you classify as manufacturing related defects, burn-in isn't a very effective method.

43.2 Steps to the HASS Process

The purpose of HASS is to develop an appropriate and cost effective manufacturing screen for your product to identify and eliminate infant mortalities. The more effective the HASS, the fewer field failures occur due to infant mortalities.

The HASS methodology consists of:

1. Write a HASS Plan
2. Choose Stresses
3. Develop a Screen Profile
4. Choose Environmental Equipment/Location of Test
5. Fixture Design/Qualification
6. Proof of Screen (POS)
7. HASS Implementation
8. Trend Analysis
9. Move from HASS to HASA—The HASA Plan
10. HASA Implementation
11. HASA Monitoring

43.2.1 Write a HASS Plan

The HASS Plan outlines the process from start to finish, including choosing the stress types and developing the screen profile, the equipment trade-off analysis, fixture design, POS, HASS implementation strategy, and the trend analysis. Be sure to tailor the plan to your specific product, rather than using a boilerplate version taken from another company or out of a textbook. The plan will serve as your roadmap, and you can use it as a decision tool during the implementation process.

43.2.2 Choose Stresses

First, review your field data to find out what types of failures have been escaping your manufacturing process. If this is your first product, check with your contract manufacturer (CM). Have them find a similar product to yours, and then have them give you information about what types of defects have been escaping their manufacturing process. With a little knowledge of the defects that have been escaping, you can then choose the right types of stresses. Many people think that more is better—for example, isn't a screen that incorporates temperature cycling and vibration together going to find more defects than a screen that only incorporates temperature cycling? It most likely is, but it will also be more costly to prove and more costly to run, and the screen may not produce a positive return on investment (ROI). A positive ROI means that the amount of money you save using the screen in the form of defects prevented is greater than the amount of money you spend on the screen itself. If you can't prove a positive ROI, you can't justify using the screen.

If the stresses you plan to use in manufacturing processes aren't the same as those used in HALT, you need to perform HALT again, but with these new types of stresses to discover the operating limits of the product under the new stresses. You can't start developing a screen until you know the operating limit of the product for each stress in the screen.

43.2.3 Develop a Screen Profile

Once you have chosen the stress types, choose the screen profile. If your screening process contains temperature and vibration, you can apply the "80/50" rule to determine the profile levels. The "80/50" rule means that you use 80% of the operating temperature discovered during HALT and 50% of the operating vibration discovered during HALT. These are both product response levels, not chamber input levels. To use 80% of the operating temperature value, take 80% of the entire temperature span and distribute evenly between the operating limits. For example, if the temperature specification for the product was 0°C to +50°C and the temperature limits in HALT were -40°C to +100°C, the temperature span is 140°C. 80% of 140°C is 112°C. The resulting screen level would be -26°C to +86°C.

However, if your upper operating margin is significantly different than your lower operating margin, then the even distribution method may not work. For example, if the temperature specification for the product was the same 0°C to +50°C and the temperature limits in HALT were -20°C to +120°C, then I recommend taking 80% of the difference between the lower specification and the lower operating limit and using this value as the lower screen profile level. Do the same on the high end. Using this method, your screen profile would be -16°C to +106°C.

From my experience, I have found that the most effective temperature screens have a minimum of a 100°C span (that can be, for example, -30°C to +70°C or -20°C to +80°C). I have also found that the most effective vibration screens have a minimum of 15 Grms of random vibration as measured on the product. If you follow the "80/50" rule and can't achieve the 100°C temperature span or the 15 Grms random vibration, then you have three options:

1. Re-HALT
2. Remove Assemblies from HASS
3. Choose a Precipitation/Detection Screen Methodology

43.2.3.1 Re-HALT

If you performed HALT early in the testing phase only and didn't go back and re-perform HALT closer to production, you should now re-HALT the product to show the improvements you made during the testing phase.

43.2.3.2 Remove Assemblies from HASS

You may have relatively low product margins as a result of one particular assembly. If the failures from this assembly during HALT were nonrelevant failures, then, if possible, remove this assembly from the system during HASS. You can either use an extension cable to bring the assembly outside of the test chamber; leave this assembly powered off; or replace the assembly with a "test assembly" (one that you will use for testing purposes only to provide the system functionality but isn't shipped to the customer). You can then substitute the real assembly back in after completing the HASS.

43.2.3.3 Choose a Precipitation/Detection Screen Methodology

If you have wide margins between your operating limit and your destruct limit, you can use a type of screen called a Precipitation/Detection Screen whereby you create a two-part profile. In the first part, the precipitation portion, develop your profile to go beyond your operating limit but within the destruct limit. As your product approaches its actual operating limit, make sure to shut it off. Then turn it back on when it comes back within the operating limit. In the second part, the detection portion, keep your profile within the operating limits. See Figure 43.1 for an example of a Precipitation/Detection Screen. Note that each portion of the screen can be one or more cycles and each portion does not have to be the same number of cycles.

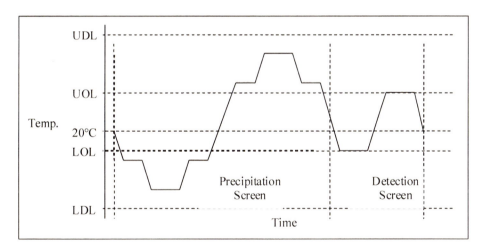

Figure 43.1: Precipitation/Detection Screen

For other stress types, choose between fifty to eighty percent of the operating limit. For example, for AC voltage margining, if your product specification is 90VAC to 130VAC, and during HALT you determined the operating limits were

80VAC to 140VAC, then you could choose a percentage of the voltage span and distribute the voltage span using the same method as we described for temperature.

Remember that whatever screen profile you choose in this step, you still need to prove that this profile is safe and effective. This comes in Section 43.2.6.1 and Section 43.2.6.2 respectively.

43.2.4 Choose Environmental Equipment/Location of Test

In order to implement HASS, you will probably require specialized environmental equipment. If your facility doesn't have this equipment, you will need to perform an ROI analysis to determine whether it makes sense to:

1. Purchase the equipment—If the ROI analysis shows that it makes sense to purchase, you will need to choose the proper piece of equipment from the many possible choices.

2. Use outside test labs—If the ROI analysis shows that it makes sense to use an outside test lab, you will need to decide which lab best fits your short-term and long-term needs.

3. Pay your contract manufacturer (CM) to take care of HASS for you (assuming they have the equipment)—If the ROI analysis shows that it makes economic sense to pay for your CM to take care of HASS for you, you need to work closely with them during the development and implementation of the screen. If they don't have experience in developing screens, you should either go to their site and develop the screen yourself on their HASS chamber or find a test lab that has the same make, model, and revision of chamber. It is important that you develop the screen on the same make, model, and revision chamber you plan to use during HASS implementation because the temperature, vibration uniformity, and vibration frequency response can vary significantly from chamber to chamber.

43.2.5 Fixture Design/Qualification

Once you identify the environmental equipment, develop and then qualify the fixture that you will use to hold the samples in the chamber. There are four steps to this process:

1. Determine the number of products that you can screen at one time to maximize throughput.

2. Design a fixture that will accommodate this throughput.

3. Build the fixture. If you don't have the capability of building your fixture, there are a number of good fixturing suppliers that can help you.

4. Qualify the fixture to ensure that it provides uniform stress across all fixture locations. If you are using temperature as a stress, you should ensure the fixture provides uniform thermal airflow across all fixture locations. If you are using vibration as a stress, you should ensure the fixture provides uniform vibration energy across all fixture locations.

Note that the fixture is likely going to be different from the HALT fixture for four reasons:

1. In HASS, you want to minimize the amount of setup and teardown between sets in order to maximize the throughput. Therefore, the fixture should have quick disconnects.
2. In HASS, you want to test as many products at a time as possible to maximize the throughput. When testing more than one product at a time, it is important that each product per screen run receive the same amount of thermal and vibration stress. Therefore, the fixture must remove variability from position to position.
3. In HASS, you want the products from one screen run to the next to receive the same amount of thermal and vibration stress. Therefore, the fixture must remove variability from one installation to the next.
4. In HASS, you will be connecting and disconnecting cables and connectors between runs, and these cables and connectors may start to wear out over time. Therefore, you should design your fixture so that the cables and connectors can be replaced easily without having to replace the entire fixture.

43.2.6 Proof of Screen (POS)

The POS ensures that the screen you have developed is both safe and effective. The POS consists of two different components:

1. Safety of Screen (to prove the screen is safe)
2. Proof of HASS Strength (to prove the screen is effective)

43.2.6.1 Safety of Screen

Using the fixture you develop, you should demonstrate that the chosen screening process leaves samples with sufficient life left in them to survive a normal lifetime of field use. You need to accomplish Safety of Screen by developing a HASS profile from the HALT results, then taking one set of products (a set consists of the number of products to be screened at one time during Production HASS) and subjecting it to multiple passes of the proposed HASS. If a failure occurs, reduce the stress levels and repeat the process. This demonstrates that the selected HASS leaves sufficient life in the product.

43.2.6.2 Proof of HASS Strength

Next, you should determine if the screen you chose is strong enough to find defects. There are four methods of Proof of HASS Strength you can use to ensure your screen is strong enough:

1. Use No Problem Found (NPF) Samples
2. Seed Samples
3. Run Higher Number of Cycles
4. Adjust Screen Based on Product Performance in the Field

43.2.6.2.1 Use No Problem Found (NPF) Samples

Use NPF samples from manufacturing or from the field (if they are available) to help confirm that the screen can find intermittent design issues. If you use field NPFs, make sure you choose samples in which your customer specifically complained of a hardware problem. In some cases, customers won't tell you why they returned the product (or they may give you incomplete information), and in these situations, the screen may not be able to find any problems.

43.2.6.2.2 Seed Samples

You can "seed" one or two samples of the product with defects to ensure that the screen is able to detect these types of defects. "Seeding" samples involves purposely inserting defects into the product. This is analogous to the process of testing software by inserting bugs and then ensuring that the test routines are able to find the bugs. This method can be effective, but it can also be inconclusive. For one thing, you may not be able to create a defect close enough to a real defect, and even if you can create a representative defect, this is only one of many different types of defects; it doesn't prove that the screen will be effective at finding other types of defects. Therefore, we don't recommend seeded samples as the only method of Proof of HASS Strength.

43.2.6.2.3 Run Higher Number of Cycles

Run a higher number of cycles of the screen at the inception of the screen, then reduce this number based on the screen results. For example, if you were planning on introducing the screen with one temperature cycle, introduce it with four cycles instead. Then, monitor performance after each cycle and plot the results. If the majority of the failures occur in the first cycle, then you can go with the one temperature cycle screen you developed. If the majority of the failures happen in the 2nd, 3rd, and 4th cycles, then you need either to stay with more cycles long-term, or expand the screen levels (or add stresses).

43.2.6.2.4 Adjust Screen Based on Product Performance in the Field

Perform Safety of Screen and do not perform Proof of HASS Strength. Then, closely monitor the performance of the product in the field and "tune" the screen (modify the screen to increase the levels or include more stresses) based on field failures that you can attribute to screen escapes. This process can take several months, and the tuning comes at the expense of your customers, so your feedback should be quick to shorten the tuning time.

If you make any changes to the screen levels using any of these four methods, you will need to re-run the Safety of Screen.

43.2.7 HASS Implementation

Once you develop the screen and choose the location of the screening, start running the screen at the designated location and collect statistics. Make sure your throughput and yield is at the level you expect. In addition, make sure you integrate the CLCA process with HASS so that when failures occur, you provide corrective action quickly and effectively.

43.2.8 Trend Analysis

You should monitor field performance and perform a trend analysis to determine if the screen is missing any infant mortality failures. If you experience field failures, then look at the screen to see if you can refine or "tune" it. If you make any changes to the screen, you will need to re-run the POS again. This is a continual process.

43.2.9 Move from HASS to HASA—The HASA Plan

If your product is going to be a high-volume product (my rule of thumb for calling something "high-volume" is when the volume reaches about 1,000 systems a month), then you should write a HASA Plan to move from 100% HASS to a HASS audit, also known as HASA.

When writing the plan, make sure to describe what criteria need to be met in order to move to HASA. The two key criteria that you need to satisfy before switching from HASS to HASA are the defect rate must be at its desired level and the process must be stable.

In the HASA plan, decide on the following parameters: detection shift level, alpha risk and beta risk levels, and sample size. The formula for sample size N is:

$$N = [(Z_\alpha + Z_\beta)2 * p * q] / D \qquad \text{(Formula 43.1)}$$

where
Z values are from a normal distribution table.
p is the baseline failure rate.
q is the success rate (1-p).

D is the detection shift level. It is the shift in percentage from your current pro-duction failure rate to a new production failure rate caused by a process change. When you move from 100% testing to sample testing, you can't detect this shift immediately. This means you will be shipping product at this higher failure rate for a period of time until you can detect this shift. The lower the detection shift value, the more samples you must test before you discover this change. For example, if you set your target failure rate at 2%, and your current production line is meeting this, then you will need to perform HASA on N samples before detecting this shift.

Alpha risk (α), sometimes referred to as a Type I Error, is the producer's risk, or the risk that you *won't* ship a good product because the data indicated that the good sample was bad. In other words, is the risk of rejecting the hypothesis that the mean hadn't changed when it hadn't (in other words, the mean didn't change but your test showed that it had changed).

Beta risk (β), sometimes referred to as a Type II Error, is the consumer's risk, or the risk that you *will* ship a defective product because the data indicated that the failed sample actually is good. In other words, is the risk of accepting the hypothesis that the mean hadn't changed when it had.

Example: If the baseline failure rate is 10% and we wish to detect a 5% shift from the baseline, and the and values are 0.20, then calculate N as:

$N= [(0.842+0.842)2 * (0.1) * (1-0.1)] / (0.05) = 102$　　　　　　*Formula 43.2*

43.2.10 HASA Implementation

Once you have a stable process and your defect rate is at its desired level, then you can move to HASA. Your production team probably has a date in mind when the volume is going to increase, as well as how fast it is going to ramp up. Make sure you set the date for the switch to HASA to align with the produc-tion needs, then closely monitor and aggressively implement corrective actions so that you meet the two criteria outlined in Section 43.2.9 at the time your schedule calls for the HASA cutover. You will run into a problem if you haven't met your criteria at this time because you can't switch over, and you won't be able to keep up with the production volume while screening 100% of the systems.

43.2.11 HASA Monitoring

You should monitor the performance of the product during HASA and make decisions when to increase/decrease the sample size. You will need to perform this ongoing statistical analysis to ensure that the HASA process is running smoothly.

43.3 When to Change a Screen

After you perform HASS Development and you begin the screening, you may never change the screen over the life of the product. A common mistake is to change a screen if failures begin to occur, even though these failures may be indicative of a process shift. However, you should still monitor the screen over the life of the product because there are times when you should change a screen. Feedback from the production line and from the field are the two key sources of information to determine if you need to change a screen. Keep a log of the screen stress levels to help with determining process drift.

Feedback from the production line is critical to ensure you analyze failures to determine if lot shifts have occurred. Compare these failures with the failures found during HASS to determine if the screen is finding the same types of failures (and hopefully more) as does the production bench testing.

Feedback from the field is equally critical because this may indicate that the screen is missing failures (too weak) or is causing latent defects (too strong). Depending on the HASS Development method used, the screen may not be strong enough to find some defects because HASS Development may have stopped short of finding the optimal limits. Conversely, the screen may be too strong and cause latent defects, especially with respect to wear-out mechanisms. HALT, HASS Development, and HASS may accelerate a wear-out mechanism but stop short of causing a hard failure. If this is the case, then the product may fail faster in the field than if no stress was applied during manufacturing.

43.4 When to Re-Prove a Screen

A good rule of thumb is to re-prove a screen if the product thermal/vibration response changes. A change in product response is usually either due to a change in the screen or a change in the fixturing. A change in the screen may be anything from a change in the dwell level, dwell time, ramp rate, or even a change in when a stress is applied in relation to the other stress(es). A change in the fixturing may be a change in the number of units being screened, orientation of the units, airflow, and even changes to the environmental equipment itself, resulting in changes in the product stress levels.

If you expect an increase in response, you should re-prove because of the possibility of damaging good hardware. If you expect a decrease in response, you should also re-prove the screen in order to determine if the screen is still able to find defects. However, if the decrease in response is small, trying to re-prove the screen will likely lead to inconclusive results.

In addition, you should re-prove a screen if you make a significant change to the design of the product, and if you can't prove that the design change does not weaken the product in any way. This is especially true if the design change involves a change in technology (i.e., solid state relays to mechanical relays or vice versa). If the change is significant, HALT should be re-run, then, based on the results, you may need to re-prove the screen levels.

43.4.1 Examples of When to Re-Prove a Screen

In one case, we used thin foam as a protective mechanism between the product and the fixture. However, the foam gave inconsistent vibration readings from one set to the next depending on the tightness of the fixture and on the cumulative amount of heat the foam had seen (the heat caused the foam to shrink, causing the vibration response to change). Our client switched to thin rubber. By doing this, the fixture increased the vibration response; therefore, we recommended that our client re-prove the screen to determine if the screen was now too strong.

In another case, we were testing two units at a time in a horizontal orientation. However, when moving operations, the screening was moved to a smaller vibration table. There was a need to have the same throughput. Because of lack of space on the table, our client changed the mounting from horizontal to vertical. This not only changed the vibration response but the airflow as well. In this instance, the fixture change significantly reduced the vibration and thermal response. Therefore, we recommended that our client re-prove the screen to determine if the screen was still strong enough to find defects.

43.5 HASS for Field Failures

Another area you can use HASS is on field returns that test out as NPF. In our experience, companies experience between twenty-five and fifty percent NPF rate for field returns. Some of these returns may be due to non-failure situations, such incorrect configurations and improper installation. However, many of these may be related to actual failures that you can't find with traditional testing methods only. In these situations, we recommend following up the traditional testing with the use of HASS, using the same profile that you used during production. If the sample passes the HASS profile, then you can ship the product back to the field. If the sample fails, then investigate to determine why your traditional testing wasn't able to find the failure.

RELIABILITY INTEGRATION: Integrating Reliability Predictions with HASS

How do you know if HASS is needed, and if so, how effective must the HASS be to meet your reliability goals? Let's say that you set a reliability goal at 99% in the first year of operation. You then perform a Reliability Prediction, and you just barely meet this goal. This prediction is accounting only for steady state failures and doesn't account for infant mortality failures. The calculations for infant mortality failures from Telcordia SR332 using their formula for the first year multiplier (FYM) factor show that no manufacturing screening at all can account for up to four times the number of failures in the first year than after the product reaches steady state. Therefore, if you are just barely making your goal *after* the product reaches steady state, you don't have a chance of meeting your goal *before* the product reaches steady state, unless you have an effective manufacturing screening program. There is a way to calculate how much HASS is needed, but the explanation is beyond the scope of this book.

CASE STUDY: Choosing the Correct Stresses

A computer company asked us to help them determine the best manufacturing screening methodology. They had been using various types of screens for years. Every time a major failure occurred during the manufacturing process, they changed the screen because they thought the screen was *inducing* the failures. We had them list the top failures across the product lines of interest. We listed those on a vertical axis. On the horizontal axis, we listed about a dozen different types of screens (including ones they tried in the past plus some they hadn't tried). We then filled out the matrix by marking an "X" in the boxes for those screens that would be effective at finding each type of failure. For those that we weren't sure about, we researched the failure modes on the internet, including past papers written on the subject. The screen that had the most "Xs" in it was deemed the most effective. Next, we determined how to implement this screen within the cost and schedule constraints. Our client ended up with a screen that was significantly more effective at finding manufacturing defects, thereby reducing the number of field failures. The new screen only cost a few dollars more per system. The net effect was that the new screen saved our client thousands of dollars per year. Figure 43.2 shows an example of the ROI we performed for our client. This example illustrates the power of an ROI Analysis, as well as the variables we recommend including in an ROI Analysis.

ASSUMPTIONS

Cost/day in labor for eng. (HASS Devel.)	$750
Cost/day in labor for tech (HASA)	$400
Cost for each board set (to be used in HASS Devel)	$900
Cost for unit (to be used in HASS Devel)	$2,000
= Units HASA'ed/year (note 1)	1500 Units
= Units Shipped/year (note 10)	15000 Units
= Units HASA'ed per screen (note 6a)	10 Units
= Units Burned-in per screen (note 6b)	50 Units
Chamber – Operator Cost per screen for HASA (note 7a)	$200
Chamber – Operator Cost per screen for Burn-In (note 7b)	$800
NRE Cost to Monitor each unit in Burn-In (note 9a)	$100
NRE Cost to Monitor each unit in HASA (note 9b)	$1,000
Cost per Defect not Found (note 2)	$500
% Defect/Year without screening (note 8)	0 %

SCREEN OPTIONS	1	2	3	4	
	HASA on boards only inside unit	Unit Burn-In Sample	Unit Burn-In 100%	No Screen	Units
Screen Parameters (initial estimate)	-50 to 75C @ 40C/min w 20G	50C, 48 Hours	50C, 48 Hours	N/A	
NRE Total	27,800	5,000	5,000		$
NRE for Fixture (note 11)	10,000				$
NRE for Test Monitor Costs	10,000	5,000	5,000		$
Cost of Consultant	3,000	0	0		$
Cost of HASS Development (Labor)	3,000				$
Time for HASS Development	4				days
Cost of HASS Development (Materials)	1,800				$
Number of Board Sets Required (note 12)	2				qty
Number of Entire Units Required	0				qty
Cost of Defects Not Found	738,000	888,750	$10,000	900,000	$
Percentage of Defects due to boards (note 14)	25	25	25	25	%
Percentage of Defects due to non-board issues (note 14)	75	75	75	75	%
Screen Effectiveness on boards tested (note 13)	90	10	10	0	%
Screen Effectiveness on boards not tested (note 4,13)	80	20	100	0	%
Screen Effectiveness on non-board issues for units tested (note 13)	0	5	10	0	%
Screen Effectiveness on non-board issues for units not tested (note 13)	0	20	100	0	%
Total Defects	1,800	1,800	1,800	1,800	qty
Total Board Defects	450	450	450	450	qty
Total Non-Board Defects	1,350	1,350	1,350	1,350	qty
Number of Board Defects Found	324	9	45	0	qty
Number of Board Defects Not Found	126	441	405	450	qty
Number of Non-Board Issues Found	0	14	135	0	qty
Number of Non-Board Defects Not Found	1,350	1,337	1,215	1,350	qty
Cost of Screening per Year	30,000	24,000	240,000	0	
Number of Units Screen	10	50	50		qty
Chamber and Operator Costs	30,000	24,000	240,000	0	$
Cost of No Screening	900,000	900,000	900,000	900,000	$
ROI in 1st Year (note 5)	$104,200	($17,750)	($155,000)	$0	
ROI in 5 Years (note 5)	$652,200	($68,750)	($755,000)	$0	
RANKING (by ROI)	1st	3rd	4th	2nd	

NOTES

Note 1 Test 10% of the units

Note 2 The cost/defect number is the most sensitive variable to the entire analysis. We need to be accurate about this

Note 3 This assumes a dedicated chamber. If not, this time will go up because of the setup and teardown fixture between customers

Note 4 Measure of how well the auditing process can find issues with samples not tested by stopping production when defects are found

Note 5 A number with parenthesis around it indicates it is a negative ROI

Note 6a This assumes a HALT/HASS chamber with capacity for 10 units at a time

Note 6b This assumes a burn-in room with capacity for 50 units at a time

Note 7a Approx 400 per day for room with operator 1-2 hours per day x 2 days = $800 for 2 day burn-in

Note 7b Cost of chamber amortized = $150 per day - $250 consumables - $400 operator = $800 per day/4 screens per day = $200 per screen

Note 8 This is taken from the Reliability Program Plan

Note 9a This is the cost to monitor each unit - likely a PC that can control a number of units at one time

Note 9b This is the cost to monitor each set of boards for HASA - likely a dedicated PC with semi-custom s/w and cables

Note 10 Forecast is for 15K in 1st year and then 20K in subsequent years. Went with 15K per year for simplification

Note 11 Approximation based on experience

Note 12 Assumes we populate two locations on the table with live samples, using max and min vib spots. Rest of table populated with dummy units

Note 13 These percentages are engineering estimates based on published information and based on the author's experience

Note 14 Percentages taken from System Reliability Prediction.

Note 15 Cost of screening = $1.25 per unit and escape rate = .75% per month

Figure 43.2: Example of HASA ROI

44 Manufacturing Metrics with ORT

An **Ongoing Reliability Test (ORT)** is used to determine the product MTBF during manufacturing to help derive a reliability figure through testing. An ORT can also point out weaknesses and failure modes that occur over time.

During the product launch, set up an ORT and rotate samples through the ORT from the manufacturing line to help monitor the projected life of the product on a continual basis and to identify potential weaknesses. In ORT, the rotation is accomplished by pulling systems from the manufacturing line, running them through the ORT process for a period of time (defined in the ORT plan), then returning them to the manufacturing line for shipment to a customer. Understanding the reliability of a product during the manufacturing process is as important as it is during the design process.

Figure 44.1 shows the Reliability "Bathtub" Curve, highlighting where ORT is most effective. Note that ORT is useful for measuring reliability only within the infant mortality region.

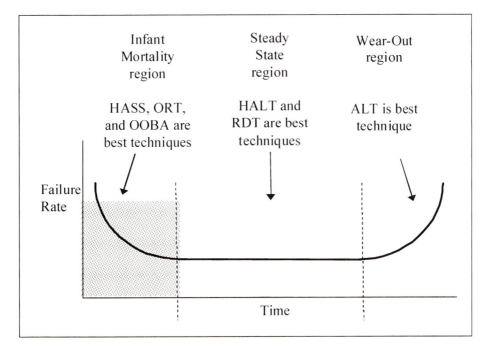

Figure 44.1: ORT Region of Reliability "Bathtub" Curve

44.1 Comparison Between ORT and RDT

ORT is similar to Reliability Demonstration Testing (RDT) except that RDT is usually performed once just prior to release of the product, whereas ORT is an ongoing test rotating in samples from the manufacturing line. Because an RDT is only performed once, it can't detect changes that occur in the reliability of a product due either to design changes or process changes. An ORT is more effective at finding these changes.

In addition, with RDT, the samples can't be shipped after completion of the test because the test may have removed a significant portion of its life. However, with ORT, you rotate the samples before appreciable wear takes place for any sample that goes through ORT. Therefore, you can ship samples that went through ORT.

Note that there is another form of ORT in which you test the product until it fails, and you don't ship the samples. You would resort to this method if the best tests to use to detect a change in reliability require taking a significant amount of life out of the samples.

44.2 Comparison Between ORT and HASA

ORT is also similar to HASA in that both techniques set out to find process changes that ultimately affect the outgoing quality. ORT is better than HASA at quantitatively predicting the reliability of the product for an on-going basis. ORT isn't as efficient or effective as HASA at finding these process changes because HASA can identify manufacturing related failures within a few hours to a few days, while ORT can take several weeks to identify the same type of defects. The faster you find the defect, the faster you can implement the change and the fewer defective samples you will ship.

44.3 ORT Process

An ORT consists of three steps:

1. Planning
2. Testing and Continual Monitor
3. Reporting

44.3.1 Planning

In the planning stage, establish a Reliability Test Goal for your production environment. This is typically expressed as a failure rate or MTBF. Then, choose an ORT acceleration model (such as the Arrhenius Model if you choose high temperature as your accelerant). Next, just like with RDT, develop an ORT Decision Matrix and outline all of the parameters that you should decide on prior to writing the plan, including types of stresses, number of samples, length of test, and confidence, along with advantages and disadvantages for each. From this matrix, decide on all of the parameters that will go into the ORT Plan. An additional variable you need to decide on is the length of time each sample will go through ORT. In RDT, this wasn't important because you typically don't ship RDT samples to a customer. However, with ORT, you will be shipping the samples that go through this process; therefore, you should make sure that you aren't removing appreciable life from the samples. Your company may even have a policy on how long you can test a product and still call it new. I recommend you check with your finance group before deciding on this parameter.

44.3.2 Testing and Continual Monitoring

In the testing and continual monitoring stage, run the ORT and continually monitor the data to determine if you are meeting your Reliability Test Goal. Analyze each failure and perform corrective action if necessary. Immediately put the samples that were corrected back into the ORT process.

44.3.3 Reporting

You should report the results of the ORT on a continual basis (typically this is done once a quarter). It is best to come up with a format to which your management agrees, then analyze the data and generate a report. If you set up the template properly at the beginning, you should be able to generate quarterly reports quickly.

RELIABILITY INTEGRATION: Integrating a Reliability Prediction with ORT

A Reliability Prediction may not be able to give an exact MTBF number, but it will give a number close enough to help determine an ORT sample size and acceleration when setting up the test. After collecting sufficient data in the ORT, compare your ORT results with your prediction so that you have this difference. You can then use this difference to develop a factor that you can use for future predictions—call it a "Prediction to ORT Factor."

CASE STUDY: ORT for a Solar Panel Company

A solar panel company purchased panels from vendors and sold them under their own label. They needed a method to gauge the ongoing reliability for each vendor, so we came up with a set of tests for their vendors to perform on a sample of the panels on an ongoing basis. We then compiled the data to show the performance of the overall population and compared this data to show how each vendor was performing relative to one another. With this method, our client was able to determine quickly when one of their vendor's processes went out of control.

45 Out-of-Box Auditing Can Catch the Easy Mistakes

An **Out of Box Audit (OOBA)** is the process of randomly taking a boxed-up system from the shipping area, opening it up, and performing an inspection and/or functional test on the sample in order to measure outgoing quality and reliability. OOBA can help your out of box defect rate if the failures are a non-functional type of failure, such as missing items from a shipment. OOBA can also help your dead on arrival (DOA) rate if the failures are a functional type of failure, such as the product doesn't power up when it arrives at your customer's site.

Note that OOBA is useful for measuring reliability only within the infant mortality region. Figure 44.1 in the previous chapter shows the Reliability "Bathtub" Curve, highlighting where OOBA is most effective.

The steps involved in an OOBA:

1. Determine Sampling Plan
2. Determine What Tests to Perform
3. Determine the Success/Failure Criteria
4. Adjust Sample Size as Needed Depending on Results of OOBA
5. Adjust Sample Size as Needed Depending on Results of Field Data

45.1 Determine Sampling Plan

With any type of audit, you should determine how many samples you plan on testing. This largely depends on the number of failures you can accept, as well as your past history.

45.2 Determine What Tests to Perform

Depending on the complexity of the testing process, you will usually just repeat all of the tests that you performed on the product the first time it went through the manufacturing process at the system level. However, in some cases, you will reduce this to testing only the critical parameters if you are concerned about specific types of failures, or if the original testing is too lengthy to repeat efficiently.

If you are also performing HASA and/or ORT, then you may want to limit the amount of functional testing in OOBA and concentrate on the non-functional areas not being covered by HASA and ORT, such as configuration, software revisions, and missing parts from shipment.

45.3 Determine the Success/Failure Criteria

You will usually set the success/failure criteria the same as in the manufacturing process. However, there may be cases where you may want to tighten or relax the criteria, depending on what you are measuring. Once you have this criteria, set the audit level accordingly. You can use Formula 43.1 to determine this.

45.4 Adjust Sample Size as Needed Depending on Results of OOBA

There is a purpose for every audit. If the audit shows successful results (few failures) during the OOBA, then it is time to relax the audit and test fewer samples. Conversely, if the audit shows areas of weakness, you will likely want to tighten the audit and test more samples.

45.5 Adjust Sample Size as Needed Depending on Results of Field Data

You should get accurate and timely results from the field to determine if the OOBA is effective. If the out of box defect rate is low, but your customers are still experiencing out of box failures, then you will need to adjust what you are checking in the OOBA, as well as the percentage you are checking.

RELIABILITY INTEGRATION: Integrating HASS with OOBA

HASS is a great technique to ensure that the product is free from infant mortalities, but what about the rest of the shipment? The manuals, cables, configuration, software revisions, and other non-functional elements may be just as important to your customers. This is one area where the OOBA process really can work well. If you have complex shipping configurations or numerous different shipping configuration combinations, you should randomly take products from the shipping area and check the configuration aspects of each shipment to ensure that the quality of your shipments meets your set quality levels. In addition, we find that with complex configurations, the system goes through a lot of handling during configuration, and this could lead to weakened assemblies. Therefore, after you complete the configuration check, you should run the system back through the HASS process to ensure that the system is defect-free.

CASE STUDY: OOBA on a Telecom Product

For a company making a voice mail product, we monitored their out of box defect rate and determined that they were experiencing too many failures. Most were non-functional types of failures, such as missing cables, incorrect software, and incorrect user manuals. We implemented an OOBA process to improve the quality level. At the same time, we studied the process to determine the root cause for each of the failures. We came to the conclusion that the configuration process was too complex—essentially, there were too many different possible configurations. We worked with the sales group to create "kits" to go along with each system. The kits contained a specified number of cables, manuals, and software. This drastically reduced the number of line items on each sales order, which in turn reduced the number of opportunities for failure. We continued to monitor the outgoing quality level and as it improved, we reduced the audit level of the OOBA.

Note that just because you get a process under control, that doesn't mean you should eliminate an audit process. It is better to continually reduce it but still have some level of audit present in case an issue returns. In this case, an issue could easily return if on a new product we don't create the same kit structure, or if a new product has many more kit options. Keeping the audit in place will identify when these types of changes adversely affect your quality levels.

46 Lessons Learned— Sounds Like School, Only Much Better

The **Lessons Learned** process consists of capturing all of the issues that occur in a program in a centralized system for all engineers to share, and then reviewing these issues prior to embarking on a new design.

A friend of mine at a large computer company told me that at his company, they only have a five-year memory on design and manufacturing fixes. Every five years, they tend to make the same mistakes over and over again. The mistake may be due to inadequate ESD protection, placing components too close to the edge of the board, or some other error. As long as that team is together, they probably won't make the same mistake. Nevertheless, they don't have a Lessons Learned database. Therefore, as soon as key members from the team move on, so does the knowledge, and then the next team has to discover the issue all over again.

The Lessons Learned process is one of the easiest techniques to understand but perhaps one of the more difficult to implement effectively. In fact, many companies we have assessed thought they did this well. However, when we asked further about where they stored the data and when and how often they reviewed it, and which teams had access to the information, it became painfully clear that they *didn't* have a good implementation of this technique.

Here are a few rules of thumb for the Lessons Learned technique:

1. Create and use one common database rather than trying to maintain a number of different databases. Companies tend to segregate what they do between different groups. For example, they may end up with a bug tracking database for software, a manufacturing database for manufacturing and a test database for hardware testing. Having separate databases will make it much more difficult to review past mistakes, especially if a mistake crossed between two groups, such as software and hardware or design and manufacturing

2. Have the database online and accessible via the internet (with adequate pass code protection to keep the information from outsiders) so that your team can access it anywhere.
3. Make sure the database is used by all. Training will be required to get everyone up to speed. Even your vendors should have access to areas of the database that pertain to their portion of the product.
4. Make sure that data is input into the database after each phase of the development, not just at the very end. People have a tendency to forget about issues over time.
5. Have a kickoff meeting at the start of a new project and plan to spend a few hours sorting though the database for Lessons Learned from the previous project.
6. Make sure you modify your Reliability Program Plan (RPP) as needed based on the results of the kickoff meeting.

RELIABILITY INTEGRATION: Integrating Goal Setting with Lessons Learned

Prior to your next Goal Setting meeting for a new product, go through your Lessons Learned database and review each entry with the team that created the entry to determine if the specific lessons you learned are relevant to this new product. Some lessons may be unique to a particular technology or process; therefore, this type of lesson may only be relevant if your new product will be using this technology or process. Other lessons are more generic and can apply to just about any type of product or process.

Once you have a list of all relevant lessons, go through your failure database to determine the number of failures caused by the mistake that led to the entry in the Lessons Learned database. For instance, if during the previous design you discovered an issue where the designer specified too loose of a tolerance on a particular component, and this error caused a one percent failure rate in the field, then note this next to that particular lesson.

Next, add up all of the failure rates. This figure represents the total percentage of improvements you can expect during your next product. This doesn't account for new technologies and new processes because you don't have experience with them. For these, you should come up with reasonable estimates for the expected failure rate.

Both of these figures—the failure rate improvement from avoiding previous mistakes and the failure rate estimated to account for new technologies and new processes—are excellent inputs into your next reliability Goal Setting meeting. You should then combine these with the other inputs to arrive at an overall reliability goal.

CASE STUDY: Multiple Databases Lead to Multiple Headaches
We were working with a mid-sized medical company that had multiple databases and multiple defect tracking systems. This was partly because they had purchased a number of companies over the years, but also because they grew their organization in a somewhat reactive mode. Whenever they discovered a problem with a particular area of their organization, they implemented a new database to track the data from that particular area. For each product type, they had a software bug tracking database, a testing database, a manufacturing yield database, and a field failure database. We worked with them first to set up a single database from which they could track all their information, then we showed them how to create reports to collect specific data on an ongoing basis to determine how their processes were performing. We helped them create a centralized failure review board (FRB) whereby the same individuals were responsible for understanding and signing off on each corrective action. We then made these individuals on the FRB responsible for the Lessons Learned database because they approved all changes to the product. With this new database, we reduced the amount of time to discover a problem from days to minutes. With the new centralized FRB, we reduced the amount of time to implement corrective actions from weeks to days, and the corrective actions had a much higher percentage of being effective the first time.

47 Let's Take a Look at Our Warranty Returns[7]

Warranty Review is the identification and prioritization of warranty performance tracking and cost reduction opportunities. This is essentially the metric check for the Warranty Analysis that you performed in the design phase.

Products with warranties incur expenses across many departments and processes within a company. Tracking, monitoring and controlling these expenses requires a common understanding of these processes and a common set of warranty metrics. Many organizations track the direct expenses or material costs but miss the hidden costs incurred in managing warranty. These may include labor, failure analysis, vendor penalties, spare part inventory, and logistic costs.

Prior to releasing a product, come up with a warranty cost model and get agreement from your finance team on the method of calculation. Install appropriate warranty tracking metrics to monitor your process and focus on expense reduction goals.

Then, after you release the product, collect all relevant warranty data and feed the information into your Warranty Analysis model to measure if you are on track with your warranty projections. If you aren't, make adjustments in your warranty reserves to cover the differences. You may also need to make adjustments to your model if you aren't capturing the true costs properly.

The warranty process should effectively tie the quality and reliability aspects of product development to production and reverse logistics. In the case of warranty, reverse logistics involve the flow of returned parts from the end user to the manufacturer for engineering analysis, repair, and recycling.

One of the goals of reverse logistics is the expeditious engineering analysis and feedback on the root causes of design and manufacturing problems. The outcome of these activities should result in an ultimate improvement in quality and reliability. In some cases, it should also determine the allocation of financial responsibilities within the supply chain (OEM, tier one, and tier two, for example).

Often, the relationship between dealer or retailer and consumer is a two-way street. The dealer or retailer distributes the product to the consumer while maintaining responsibilities for handling customer claims and moving them back through the supply chain. In other scenarios, the manufacturer could receive the returned parts directly from the customer, effectively bypassing the dealer or retailer, which would not normally affect the rest of the flow.

In many organizations, the customer service department of either the dealer or the manufacturer would process the claims by handling returned parts and making some initial decisions, such as sending the part for an engineering analysis or accepting or rejecting a customer claim on its merits.

In many cases, engineering analysis should determine failure modes and failure mechanisms and should trigger the appropriate actions, such as recommending specific measures to contain the problem or implementing suggestions to improve the design and manufacturing process.

Depending on the complexity of the returned parts, the engineering tasks can be accomplished by failure analysis, structured problem solving, or using a combination of existing continuous improvement tools.

Engineering analysis, for example, can determine that a part failed due to an assembly problem, end-user abuse, software malfunction, component failure, corrosion, overheating or vibration. Consequently, some of these failure modes would trigger corrective actions.

Engineering analysis findings also serve to resolve disputes between OEMs and suppliers responsible or liable for a failure. In high-volume production (and, consequently, higher-volume returns), you can't analyze every returned part; therefore, you must devise an efficient sampling strategy.

The strategy could include random sampling of returned parts or a stratified sampling, in which the first several months' worth of returns are sampled more often for new model production. In this case, a stratified sampling would help evaluate the process as quickly as possible and allow you to make timely adjustments when necessary. Additionally, trend analysis of the initial batch of warranty data can also serve as an early warning system to trigger timely corrective actions to avoid more serious consequences.

The warranty process also crosses into accounting and finance to augment the warranty accounting system by adding the information about disputed claims and to settle the remaining payables and receivables.

Engineering feedback from the field is essential to successful product design and development. You need to adjust your new product development process based on the engineering analysis of returned parts to prevent old problems from recurring in new products. This is an important part of the continuous improvement process.

A strong, flexible, and user-friendly failure reporting, analysis and corrective action system is essential to storing warranty data and using it as a source for future design improvements. It is no accident that the market for software tools focused on warranty data storage, analysis, and management has been steadily growing during the past five to ten years.

In addition, warranty data are routinely used for reliability and warranty prediction in new product development. For example, field failure data of existing products can be more accurate reliability predictors of future products than some of the traditional methods, such as reliability growth models or standards-based predictions.

Warranty analysis data also carry valuable information for supplier evaluation, dispute resolution, technology evaluation and improvement, data mining, and fraud detection and prevention. Most warranty professionals appreciate the amount of engineering and business data available in warranty claims records, but they admit that this wealth of information often remains underused due to lack of time, skills, or other resources.

RELIABILITY INTEGRATION: Integrating Warranty Analysis with Warranty Returns
Just as with other metrics-based techniques, once you establish which metrics to capture, you should then review the data that you collected to ensure that you are meeting your goals. The same is true for warranty. Set your warranty metrics during the warranty analysis, then monitor the returns to ensure that you are meeting your warranty goals. If you aren't, set actions in place to correct the issue to get back on track, or change your warranty goals for the next product.

CASE STUDY: Tracking Warranty Return Rate

A test and measurement instrument company producing handheld to bench-top precision test equipment tracked their overall product warranty rate. They knew that the larger, more complex products had more costs than the robust handheld units, yet the metric they used and their method of compiling the data didn't reveal the differences. We helped them refocus the data collection to product line categories and set up metrics to measure failure rates and warranty cost per system. The variability of the high-end products became the clear opportunity for improvement. We then worked with the management team to establish warranty projections for each product based on existing performance and business needs. Using these projections, they were able to predict their product warranty rate much more accurately.

---

7. Quality Progress Magazine, May 2010, "Discussion Warranted", Andre Kleyner, pp 25–27.

48 Management Needs a Report

Field Data Tracking is the process of collecting product performance data from the field and analyzing it by using statistics to determine trends and to measure your reliability against your original goals. **Reliability Performance Reporting** is a method of reporting back the results of how you are doing against your plan. In this report, you should capture:

1. How you are doing against your reliability goals and your schedule to meet your goals.
2. How well you are integrating each technique together.
3. What modifications you may need to make to your plan.

In the report, you can also add information on specific issues, progress on failure analyses for any failures you discover on the program, and Pareto charts and trend charts. Once you start shipping a product, there is a great deal of information you can collect to measure your product's performance. A Reliability Performance Report is the perfect vehicle for doing this. You can compare reliability performance results with your Reliability Program Plan (RPP), and make initial analytical assessments to determine how you are performing against your goals. Collecting this information on a regular basis and producing a report for your senior management is an effective method to help inform the proper people regarding the performance of products, trends, issues, warranty changes, and resolutions. The steps involved in this:

1. Collect the Data
2. Analyze the Data
3. Determine How Often to Report
4. Develop Report for Senior Management

48.1 Collect the Data

When collecting the data, gather the following information (you should set this up in advance with your field support personnel, and you may need to generate failure tags and hand them out):

1. The date/time for which the sample failed so that you can determine the time to failure data (how long between the last failure and this current failure). An on-board hour meter on your system helps with the time-to-failure data.
2. What failed and what were the symptoms?
3. Was it repeatable?
4. What was replaced?
5. How was the failure fixed?
6. Serial number of the failed assembly and the serial number of the system it came from.

48.2 Analyze the Data

Take all of the time-to-failure data and plot it on log-log paper. With enough data points, you can determine if there was one dominant failure mechanism or several, and you can also determine what would be the reliability distribution that best fits the data. Some of the most common distributions engineers use to evaluate the reliability data are Exponential Distribution, Log-Normal Distribution, Normal Distribution, and Weibull Distribution. If the data fits a Weibull distribution, then you should determine if the failures are indicative of infant mortality failures, steady state failures, or wear-out failures. You may not be able to start the failure analysis process without first knowing this.

48.3 Determine How Often to Report

You should develop reliability performance reports about once per month during the testing and initial manufacturing build. You can drop down to once per quarter after your processes have stabilized.

48.4 Develop Report for Senior Management

A Reliability Performance Report consists of:

1. Starting the report with a management summary, which includes the current reliability metrics and how they compare to your reliability goals.
2. Outlining which metrics you are monitoring.

3. Identifying which reliability techniques you are using and evaluate each technique's effectiveness and how each contributes to the goals.

4. Comparing field results.

5. Describing changes you made in the design, manufacturing process, or field maintenance to align the current reliability closer with your goals.

6. Providing backup data and analysis.

7. Describing future actions planned.

RELIABILITY INTEGRATION: Integrating Your Reliability Performance Report with Your Reliability Plan

The information in your Reliability Performance Report will point out areas of your reliability program that are working, as well as opportunities for improvement. Both pieces of information are excellent input into your Reliability Program Plan (RPP) for your next product. For example, if you experience a trend of failures in the field that were related to wear-out of a mechanical component, review your current RPP and the techniques you called out in the plan to understand which techniques should have caught the failures and why they didn't. Perhaps you failed to identify this failure mode in your FMEA, and therefore you didn't include an ALT for the component in your test plan. If this occurs, make a change to your FMEA process to add a brainstorming exercise for wear-out mechanisms. You can then integrate those changes into your RPP for your next product by emphasizing the importance of FMEA and allocating a portion of your reliability goal to wear-out mechanisms.

CASE STUDY: Developing Your First Reliability Performance Report

A consumer electronics company wasn't collecting any field data and wasn't running any type of reporting of information. However, increasing customer complaints were pressuring them to do so. We helped them set up a method of tracking their data and developed an easy method of acquiring simple statistics and reporting these statistics in a clear and concise manner. This worked for several months, after which people stopped paying any attention to the report. Soon the reports merely served as a box to check in the review process. We reviewed the way they presented their data and found that the management summary became weak and had no takeaways or concrete actions. We helped them restructure this section to get management to start paying attention to the report and to get management involved to help drive corrective actions.

49 Statistics—More Than a Four-Letter Word

Good use of **Reliability Statistics** can go a long way in product develop-ment as well as in field data analysis. Whether you use them to determine the sample size in a reliability test, to set up a DOE for making designs less susceptible to variation, or to perform a Reliability Prediction and assess the confidence level of the prediction results, the use of Reliability Statistics is powerful. Most situations don't even require more than basic use of standard deviations or the application of the Weibull Distribution.

49.1 Weibull Distribution

The Weibull Distribution is a failure distribution used to model time to failure, time to repair, and material strength. The Weibull Distribution can be expressed as:

$$R(t) = e^{-[(t-\eta)/\gamma]^{\beta}}$$ *Formula 49.1*

where:

β (beta) is the shape parameter

η (eta) is the scale parameter (how wide is the distribution)

γ (gamma) is the location parameter (if you ship a product to a warehouse and it gets installed at some later time, you can call that elapsed time ? and that time gets added to the time to failure)

Next, you can use this distribution to help describe your product's reliability through its product life cycle.

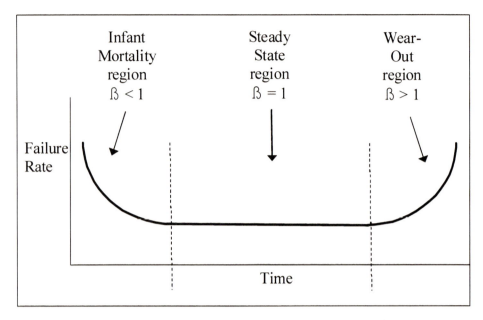

Figure 49.1: Reliability "Bathtub" Curve with Weibull Beta Parameters

As I stated at the beginning of this book, I purposely avoided use a significant amount of theory and formulas. I believe one fundamental mistake authors make when writing reliability textbooks is that they use so much statistics that they "lose their audience in the math." This tends to scare many design engineers into thinking that reliability is "for the reliability engineer." Now don't get me wrong—to understand all of the nuances of reliability well, you need to have a firm grasp on statistics. You need to know the difference between Weibull Distributions versus Normal Distributions versus Exponential Distributions, and you need to understand Bayesian Theories and how to apply Markov Analysis. However, when implementing DFR techniques, most of this knowledge isn't necessary; therefore, I will stop the discussion here on statistics.

RELIABILITY INTEGRATION: Integrating Statistics into Field Data Analysis

When you review your field data, what units do you use to describe the results? Hours? Years? Do you immediately think of MTBF when you think of field data? MTBF can be a very misleading number when trying to describe a field population because it only tells part of the story. MTBF is a timeless figure. You don't know if the population is made up mostly of infant mortality events, mostly steady state events, or mostly wear-out events. Worse, it could be bimodal, in which case you may have a few different failure modes going on. This is where statistics comes into play. You need to plot your failure data over time to get the true picture of your product reliability in the field. The same holds true for when you are setting a goal or performing a Reliability Prediction. With a goal statement, you need to budget for each time period of the Reliability "Bathtub" Curve—infant mortality, steady state, or wear-out. With a Reliability Prediction, you need to analyze the contribution from each time region.

CASE STUDY: Using Statistics During an ALT

Our client was a manufacturer of an electromechanical product, and his product was failing in the field sooner than his ALT showed. We reviewed his field data and his ALT data. We plotted the field data on log-log graph paper and determined that there existed a dominant wear-out mechanism. We plotted the ALT data on the same graph and didn't find the same wear-out mechanism. Then we reviewed his ALT Plan and discovered that our client set up an RDT instead of an ALT. He substituted samples for test time and never actually took a single sample to failure. Instead, he tested for a pre-determined amount of time, then stopped the test, then added up the total test time. This is the style of testing you would use if you expected that the failure was a steady state failure event. However, for wear-out failures, you should take the samples to failure (or utilize the method of "Leading Indicators" discussed in Section 36.3, whereby you take enough data to observe a change in performance characteristic of the onset of a failure, then extrapolate the time to failure). We showed our client how to set up a proper ALT, take a number of samples to failure, then plot the failures on a the graph. With this method, our client was able to predict much more accurately how long his product would last before wearing out.

50 When Faced With Obsolete Parts, Turn To EDA

Electronic Design Automation (EDA) is a category of software tools for designing electronic systems, such as printed circuit boards and integrated circuits. Several classes of challenges routinely arise in ongoing programs in industries with lengthy life cycles, such as military, aerospace, telecommunication, medical, and energy. Unlike the fast-paced commercial consumer world, such programs often have significant non-recurring engineering (NRE) costs, and they need long-term life usage to recover the costs invested. Military and medical programs have potentially severe impact on users when they don't work as intended. In this setting, the use of EDA techniques throughout the product life cycle (PLC) offers an improved way to address issues in parts obsolescence, pattern failure emergences, parts-counterfeiting situations, qualification of commercial off-the-shelf (COTS) products, and agency requirements on design changes.

Design engineers have used EDA tools to design most circuitry in use today. Generations of EDA techniques have been used in engineering departments of countless companies, and an entire industry has developed around EDA to supply and support their use by these companies. Most electrical engineering students in the last two decades have been introduced to these techniques in school, and they have a general familiarity with the usage of EDA tools.

As the capabilities of EDA tools have evolved, the thrust of the EDA firms has been toward higher performance and more functionality. However, along with this comes a higher price per license. Therefore, many companies restrict the usage of EDA tools within the design functions. Other parts of your organizations that are active in the product life cycle often have no access to these tools. For example, we rarely see engineers using EDA tools for Derating Analysis and FMEAs. Here are a few other ways where engineers can use EDA tools:

1. Detailed assessment of pattern failure recurrences
2. Circuit tolerance issues
3. Electromagnetic interference (EMI) properties
4. Power dissipation and dynamic signal conditions

All of these are additional ways to make effective use of your existing design files. These are expansions that bring more of your company's professional talent in contact with the design modeling and performance characterization. It allows a more detailed use of existing design files and lowers your company's dependence on the original designers, thereby freeing them from having to support completed designs if field issues arise. Thus, analysis software and design files can also serve well during the rest of your PLC.

Another part of your PLC where EDA can really help is during production. Engineers using production programs are routinely surprised by unbudgeted incidents, such as when:

1. Critical parameters have changed
2. Designs evolve over revisions
3. Tolerance windows have shifted
4. Program-specific waivers or deviations have taken place

This is compounded by the fact that you typically have no technical expertise available, either because the original designers have left the company or they aren't available to support these incidents. If you are faced with this on your production program, you can employ EDA to mitigate the issues quickly.

A good EDA technique for PLC is the tool Simulation Program with Integrated Circuit Emphasis (SPICE) because it is well known throughout the industry. For virtually every EDA package in design use, there are compatible versions to support PLC uses. Quite often the same EDA vendor will have an offering or a license modification.

RELIABILITY INTEGRATION: Integrating EDA in Manufacturing with EDA in Design

When faced with an obsolescence issue or when adding a new vendor's component to your approved vendor list (AVL), go to the EDA files already created during the design phase and tweak the parameters for a quick determination whether the new component will work as well as the previous component. If the analysis shows the component isn't compatible, you can rule out this component without doing any further analysis or testing. If the analysis shows the component is compatible, then you can move to the next step of the qualification, which is to install the component into the circuit and run it through the qualification testing.

CASE STUDY: EDA—Accessible Circuit/System Dynamics to Consider

A military equipment company had a number of components on their product that were going obsolete at the same time. We used the EDA techniques introduced in this chapter to evaluate the following areas:

1. Impact of capacitor equivalent series resistance (ESR) changes in switching regulators,
2. Changes in transient response, circuit stability, or electromagnetic interference (EMI) issues, and
3. Compromises in performance windows, derating, or end-of-life (EOL) margins.

They also decided to perform a design update to add a few new features and to reduce the power consumption as one of these new features. We used the same EDA methodology and evaluated the product for possible effects of the new features on the rest of the design, as well as possible effects of the power dissipation reduction on the rest of the design.

The use of EDA saved our client weeks of design analysis time.

Part VI
Summary

CONCEPT
DESIGN
PROTOTYPE
MANUFACTURING

RELIABILITY INTEGRATION ℠

Reliability Engineering Services Integrated Throughout the Product Life Cycle

Recap of Book

The book you have just finished is a practical guide based on my twenty-five years of experience in working with over 500 different companies in over 100 different industries.

We have seen best in class companies, worst in class companies, and everything in between. Through all of this, we learned that there is no magic recipe for reliability. There never has been, and there never will be. Reliability is all about identifying risks and then mitigating them so that they don't become issues in your customers' hands.

In Part 1, we introduced the topic of Design for Reliability (DFR) and discussed the importance of Reliability Integration, a theme we brought up in every chapter of the book with every technique we covered.

In Part 2, we described different reliability planning techniques you can use during the concept phase of your product life cycle to develop a reliability goal and write a Reliability Program Plan (RPP).

In Part 3, we described different reliability analysis techniques you can use during the design phase of your product life cycle. In this phase, your product design is still on paper, so it is much quicker and less expensive to make changes to your design. FMEAs, Reliability Predictions, and Derating Analysis are some of the most effective techniques in this phase.

In Part 4, we described different reliability testing techniques you can use during the prototype phase of your product life cycle. In this phase, you are testing prototypes of your product. The earlier in this phase you can find issues, the less expensive the change will be and the lower the chance of impacting your development schedule. HALT, RDT, and ALT are some of the most effective techniques in this phase.

In Part 5, we described different reliability testing and analysis techniques that you can use during the manufacturing phase of your product life cycle. In this phase, you are testing samples of the product that will be shipped to your customer. Therefore, you should develop testing methods that are both safe and effective. HASS, ORT, and OOBA are some of the most effective techniques in this phase.

See Figure S.1 for a pictorial view of how each reliability technique fits into an overall reliability program.

We hope you found this book useful, and I look forward to hearing from you to learn your favorite techniques and best stories. I will leave you with one final case study that best summarizes this book and the techniques we have described.

CASE STUDY: Using the Methods in this Book to Develop a Reliable Product

We were working with a medical company on developing and executing a reliability program. At the beginning of the development program, we facilitated a goal-setting meeting, then helped our client write a Reliability Program Plan (RPP). We then worked with them through the execution of each phase of the plan, providing both formal training and one-on-one coaching. At the end of the program, after they had been shipping their product for a few months, we asked them how their product was performing. They responded, "We haven't had any failures." "That's great," we replied. "No, you don't understand," said our client. "Prior to your help, we had never released a product that didn't have a failure in the first few months." The moral of this story is that the methods we describe in this book really do work. Develop a goal, write a plan, and execute the plan. It's that simple!

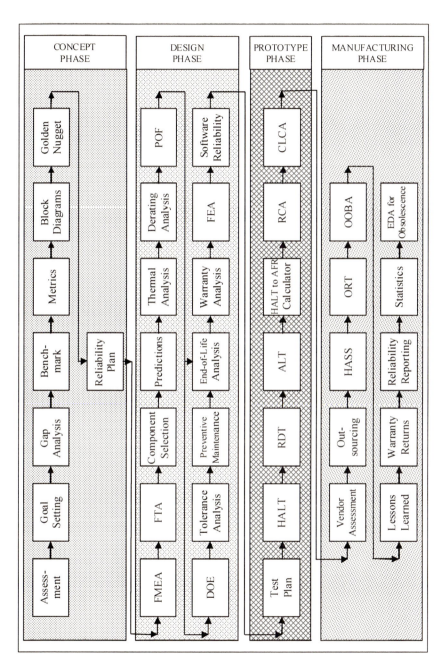

Figure S.1: Diagram of Reliability Techniques

A Software Reliability Growth

The following section of the book on Software Reliability Growth was provided by Mark Turner of Enecsys. This is referenced from Section 31.3.5 from the main section of this book.

You can measure and manage the reliability growth of any software (or even hardware) development using an appropriate model, of which many exist, some of which are more suitable than others. Perhaps the most suitable for software development is the Rayleigh model.

Software design and development is a continuous process where you provide functionality using source code. Unfortunately, despite the best intentions of engineers, you may introduce defects as you create source code. Therefore, you will benefit from modeling the creation, identification, and elimination of code defects as a function of time.

Throughout the software development process, there are numerous opportunities for you to introduce defects. We provide a typical Rayleigh curve in Figure A.1, which illustrates the defect insertion and discovery process. This shows how you can identify and address defects that you introduce at an earlier phase in the software development. There will come a point in any software development program where you maximize the defect discovery rate, after which you reduce over time the quantity of remaining defects. Here the leading bar graph illustrates code defect insertion, which can often begin at the start of the development project. Because code defects are often related to the amount of engineering effort, the rate at which you introduce them often is directly proportional to that effort.

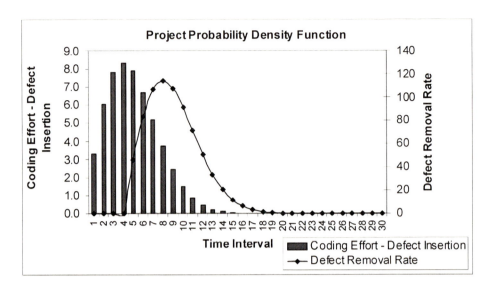

Figure A.1: Rayleigh Estimation Model with Effort and Defect Curves

The lagging curve illustrates the defect removal rate, with problems being addressed at a later date than when you originally inserted them, which can hinder project progress and negatively impact customer satisfaction. You can partially mitigate this by conducting design reviews, code inspections, and early module testing, as these activities will often assist you in discovering inserted defects as early as possible, thus moving the defect discovery and correction curve to the left.

Eventually, the quantity of defects still present in the code will equate to the original reliability target, and as you discover and address further defects, the Software Reliability increases, or grows. You can manage the rate at which you address defects by setting software defect targets. This has to begin by estimating how many defects are likely to occur, then addressing those defects by implementing a Software Reliability growth management program in which you plan and schedule the necessary resource to ensure you achieve your reliability target.

A.1 Implementing the Rayleigh Model

You can use the Rayleigh function to forecast the rate at which you identify defects during the software development program as a function of time. It is a specific instance of one of the models in the Weibull family of reliability models.

Appendix A: Software Reliability Growth

This model is particularly suited to Software Reliability modeling as it provides a good representation of the vector sum of a large number of random sources of defects, none of which dominate. The Rayleigh model provides an effective iterative design process in which feedback is inherently part of the solution process, and in fact it closely approximates the actual profile of defect data that you collect during software development programs.

Monitoring software development defect metrics can provide you valuable input into planning engineering and Root Cause Analysis (RCA) efforts, and it helps you to quantify the maturity of the software you are developing. Collecting defect metrics related to engineering effort, project duration, and type over several development projects provides a great opportunity to analyze trends, which can then provide you with more accurate resource predictions for new projects. If you lack such trend data (which typically is a problem when you first deploy the Rayleigh model), then you may have to use industry data as an alternative guide. While this alternative approach may not factor in the abilities of your actual design team, it does at least provide a reasonable estimation to begin with.

After your organization completes multiple projects, you will benefit from reviewing predicted versus actual defect counts, as this enables you to refine the original estimates and improve the model for future development projects. As the project progresses, compare the initial defect count estimates with the quantity of defects you actually address. If you find that the actual defect count is significantly higher than predicted, then the model has generated an early indication that a significant problem may exist. On the other hand, if you find that the actual defect count is significantly less than the initial prediction, then you should confirm that the identification process is indeed sufficient to detect the anticipated defects. Once you confirm this, then you can conclude that your defect insertion rate is actually less than predicted.

In using the Rayleigh model, you should determine the parameters for the total anticipated engineering effort, the total number of defects that you expect to insert into the code, and the time period to reach the peak estimate. Knowing these parameters will enable you to plot the cumulative distribution function (CDF). We've shown an example of a CDF Chart in Figure A.2 and a CDF Table in Figure A.3. The Rayleigh model parameters of Figure A.3 are 55 man-months of engineering effort, 755 inserted defects, and 4 months to reach the peak estimate. For the defect plot, you must define an additional value associated with the estimated lag behind the start of the project effort to account for defect detection and correction effort, which in Figure A.3 is 4 months.

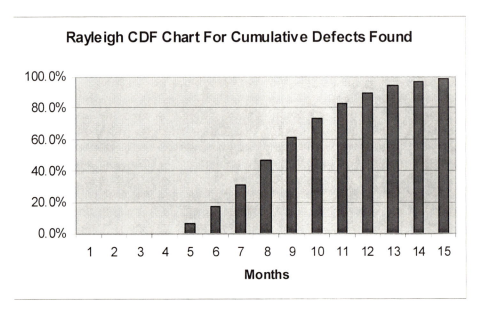

Rayleigh CDF Chart For Cumulative Defects Found

Figure A.2: Rayleigh Cumulative Distribution Function (CDF) Chart

Figures A.2 and A.3 illustrate the relationship between the project effort and the number of defects that you insert into the code, which enables you to make decisions regarding the impact that any code changes are likely to have and in changes to the code delivery date. From this example, you can conclude that a delivery schedule of eight months would be completely unrealistic, as a significant number of defects will still be present in your code, whereas a delivery schedule of twelve to fifteen months is more realistic. Delivery schedules in between require you to make a schedule versus reliability tradeoff.

Month	Effort	Inserted Defects	Cumulative defects discovered	CDF
1	3.3	0	0	0.0%
2	6.1	0	0	0.0%
3	7.8	0	0	0.0%
4	8.3	0	0	0.0%
5	7.9	79	79	10.6%
6	6.7	134	214	28.6%
7	5.2	153	366	49.0%
8	3.7	138	504	67.4%
9	2.5	105	609	81.4%
10	1.5	68	677	90.5%
11	0.9	39	716	95.7%
12	0.5	19	735	98.2%
13	0.2	8	743	99.4%
14	0.1	3	746	99.8%
15	0.0	1	748	99.9%
Total	55	748		

Figure A.3: Rayleigh CDF Table

However, if early delivery is unavoidable, the CDF can aid in planning reliability growth activities and managing customer expectations where multiple deliveries are viable.

B Summary of HALT Results at a HALT Lab

This is referenced from Section 34.2 from the main portion of this book.

This study analyzed HALT data from many different companies from a variety of industries and illustrates the following concepts:

1. HALT can be applied to a wide variety of electrical and electromechanical products in almost any industry.
2. Products today are much more robust than in the past and therefore these methods are necessary to improve product reliability.
3. Random vibration is much more effective than rapid temperature cycling for accelerating defects, and the combined environment of random vibration and rapid temperature cycling is even more effective.
4. The types of failures that HALT discovers are the same types of failures that are found in the field.
5. Don't overdesign your product—design to your specifications, then let HALT show you were the weaknesses are.
6. Your product is likely more robust than you think. Typically, there are just a few issues you need to concern yourself with, and HALT is an excellent technique to find these issues.

The examples in this study were obtained between 1995 and 2010.

I subjected each product in the study to a standard HALT regime of cold step stress, hot step stress, rapid temperature transitions, vibration step stress, and combined rapid temperature transitions with vibration step stress. For each product, we used either a Qualmark or a Chart HALT chamber, similar to the ones shown in Figure B.1.

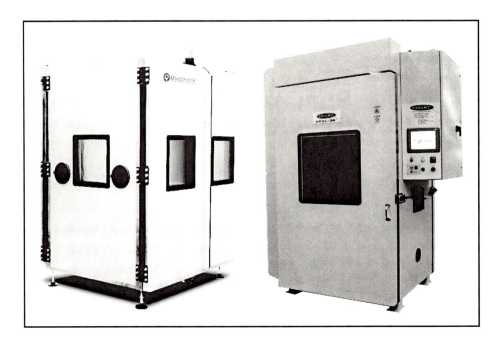

*Figure B.1: Qualmark OVS2.5 HALT Chamber (left) and Chart Real36
(right) Used for Study*

The study is comprised of data on 294 products from 157 companies across thirty-eight different industries. The majority of products were electrical, but several of the products had mechanical components as well. No specific data about any one product is presented so that confidentiality is maintained between the lab and its customers. The industries participating are shown in Table B.1. This is an updated study from the one I performed in 1996. I reviewed more data from the past fourteen years and added to the study as shown in Tables B.1 and B.2. I reviewed the new results, compared them to the old results, and found that they were very similar. Therefore, the takeaways are the same. For this reason, I did not update with the new data for Figure B.2 and for Tables B.3 through B.10.

Table B.1:   Distribution of Companies by Industry

	Industry Types	Number of Companies	Product Type
1	Networking Equipment	26	Electrical
2	Medical Devices	25	Electromechanical
3	Defense Electronics	11	Electrical
4	Microwave Equipment	6	Electrical
5	Computers	6	Electrical
6	Semiconductor Capital Equipment	6	Electromechanical
7	Video Processing Equipment	5	Electrical
8	GPS	4	Electromechanical
9	Robotics	4	Electromechanical
10	Smart Meter	4	Electrical
11	Commercial Aviation Electronics	3	Electrical
12	Display	3	Electrical
13	Oil and Gas	3	Electromechanical
14	Solar	3	Electrical
15	Storage System	3	Electromechanical
16	Homeland Security	2	Electromechanical
17	Automotive	2	Electromechanical
18	Computer Peripherals	2	Electromechanical
19	EDA Platforms	2	Electrical
20	Electric Car	2	Electromechanical
21	Fiber Optics	2	Electrical
22	Hand-held Equipment	2	Electromechanical
23	Industrial Applications	2	Electromechanical
24	LED/Lighting	2	Electrical
25	Optical	2	Electromechanical
26	Point of Sales Equipment	2	Electromechanical
27	Audio	2	Electrical
28	Recreation	2	Electromechanical
29	Government/Research	2	Electrical
30	Restaurant/Food Industry	2	Electromechanical
31	Space	2	Electrical
32	Audio	2	Electromechanical
33	Supercomputers	2	Electrical
34	Teleconferencing Equipment	2	Electromechanical
35	Test Equipment	2	Electrical
36	Wireless	2	Electrical
37	Mobile Phone	2	Electromechanical
38	Nuclear	1	Electromechanical
	TOTAL	157	

The majority of companies included in this data have products that are intended for the office environment. The product's end-use environments are shown in Table B.2.

Table B.2: Distribution of Products by Environment Type

Environment Type	Percent of Total	Thermal Environment (1)	Vibration Environment (1)
Office	38%	0 to 40°C	Little or no vibration
Office with User	17%	0 to 40°C	Vibration only from user of equipment
Vehicle	17%	-40 to +75°C	1–2 Grms vibration, 0–200 Hz frequency
Field	15%	-40 to +60°C	Little or no vibration
Field with User	9%	-40 to +60°C	Vibration only from user of equipment
Airplane	4%	-40 to +75°C	1–2 Grms vibration, 0–500 Hz frequency

Note 1 All values are approximates based on a combination of data from individual customers and from Telcordia and military specifications.

B.1 Results of Study

I combined and summarized the results for all of the products tested. The HALT methodology I used was similar for each of the products represented—the average time to complete a HALT was three to four days. The majority of customers completed the HALT in one visit, correcting the problems at their facility after the end of the HALT. Some, however, decided to divide the HALT into two or more visits, verifying corrective actions before implementing them on production versions of the product.

The failure percentage by stress type is shown in Figure B.2. The order we applied the stresses for each product was

1. Cold step stress
2. Hot step stress
3. Rapid temperature transitions
4. Vibration step stress and
5. Combined environment (consisting of vibration step stress combined with rapid temperature transitions)

Because we subjected all of the products to the stressing in this order, it becomes apparent when reviewing Figure B.2 why the failure distribution occurred as it did. For instance, because we always performed cold and hot step stress prior to rapid temperature transitions, a much higher percentage of failures occurred during the step stress. Our clients corrected these issues or created workarounds for these issues prior to performing rapid temperature transitions. Likewise, because we always performed rapid temperature transitions and vibration step stress prior to performing combined environment, a much higher percentage of failures occurred during the rapid temperature transitions and vibration step stress. Our clients corrected these issues (or created workarounds for these issues) prior to performing combined environment. Nevertheless, note that it is important to perform combined environment because had we skipped this stress, we would have missed twenty percent of the potential failures.

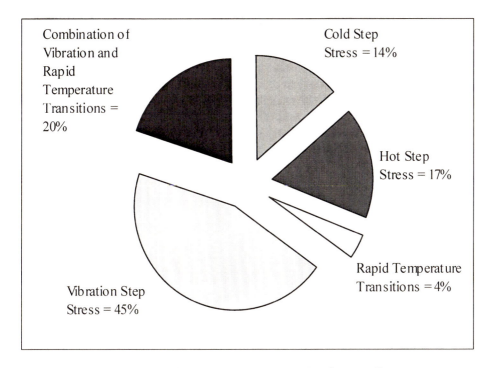

Figure B.2: Failure Percentage by Stress Type

In Tables B.3, B.4, and B.5, I summarize the HALT limits and present them in three different ways. For Table B.3, I grouped all of the products together, and I show the HALT Limits for the entire set of products. For each limit, I give the average, most robust, least robust, and median values. For Table B.4, I grouped the products by environment using the same environment categories as in Table B.2. For Table B.5, I grouped the products by product application using the three product applications of military, commercial, and field.

For all products, I used a HALT chamber with input vibration consisting of broadband energy from 0 to 10 kHz. This was a key factor in the types of failure modes uncovered because the products had a wide range of component sizes and technologies, and each required different frequency bands for excitation, i.e., large mass parts require low frequency (10 Hz to 1 kHz) to excite them while small mass parts require high frequency (> 1 kHz). In addition, the HALT chamber had a temperature rate of change of 60°C/minute which we used for rapid temperature transitions and combined environment stressing. This was also important because we know some of the failures we uncovered were due to the temperature rate of change.

For the temperature data, I recorded the thermal data in °C and used the average product temperature at the point of failure for each product tested as the measurement point. For the vibration data, I recorded the data in Grms in a bandwidth from 0 Hz to 3 kHz (even though the system was providing energy up to 10 kHz). I took the measurement up to 3 kHz because this is traditionally the bandwidth that HALT practitioners have used and is a measure of the maximum product response level at the point of failure for each product tested. For each individual reading in each of the averages, the data point I used was the worst case data point if I tested more than one product and was the limit of the product before our clients implemented any corrective actions. For all vibration readings, I used product response values and not table inputs.

Table B.3: HALT Limits by Limit Attribute

	Thermal Data, °C				Vibration Data, Grms	
Attribute	LOL	LDL	UOL	UDL	VOL	VDL
Average	-55	-73	+93	+107	61	65
Most Robust	-100	-100	+200	+200	215	215
Least Robust	15	-20	+40	+40	5	20
Median	-55	-80	+90	+110	50	52

The results shown in Table B.3 are significant because they indicate that even though the majority of the products were commercial products using commercial grade components for the office environments, the average limits achieved were far beyond the limits specified by the component manufacturers. In fact, the majority of the components used in commercial products are rated to operate from 0°C to +70°C with little or no vibration being specified.

Table B.4: HALT Limits by Environment

Environment	Thermal Data, °C				Vibration Data, Grms	
	LOL	LDL	UOL	UDL	VOL	VDL
Office	-62	-80	92	118	46	52
Office with User	-21	-50	67	76	32	36
Vehicle	-69	-78	116	123	121	124
Field	-66	-81	106	124	66	69
Field with User	-49	-68	81	106	62	62
Airplane	-60	-90	110	110	18	29

The results shown in Table B.4 follow closely with the expected order of limits (although the absolute levels are probably beyond expectations).

Table B.5: HALT Limits by Product Application

Product Application	Thermal Data, °C				Vibration Data, Grms	
	LOL	LDL	UOL	UDL	VOL	VDL
Military	-69	-78	116	123	121	124
Field	-57	-74	94	115	64	66
Commercial	-48	-73	90	95	32	39

The results shown in Table B.5 again follow the expected order of limits closely, with the military products being the most robust. However, notice that the temperature limits achieved for commercial products was close to the temperature limits achieved for military products. This is another strong argument for the use of commercial-off-the-shelf (COTS) components for military products because the commercial components are almost as robust, usually providing ample margins even in the military application, and commercial components are almost always significantly less expensive than military components.

B.2 Failure Summaries by Stress Type

Tables B.6 through B.10 are failure summaries for each type of stress applied. The significance of the data in these tables is that the majority of failures shown are common field failures for commercial equipment. The HALT process merely accelerates what will most likely take place over a much longer period of time under field usage.

One way to determine if HALT will be effective in your application is to compare Tables B.6 through B.10 with your field failures. To the extent they match up is likely to be how effective HALT will be for you.

Table B.6:   Cold Step Stress Failures

Failure Mode	Qty
Failed component	9
Circuit design issue	3
Two samples had much different limits	3
Intermittent component	1

Table B.7:   Hot Step Stress Failures

Failure Mode	Qty
Failed component	11
Circuit design issue	4
Degraded component	2
Warped cover	1

Table B.8:   Rapid Temperature Transition Failures

Failure Mode	Qty
Cracked component	1
Intermittent component	1
Failed component, cause unknown	1
Connector separated from board	1

Table B.9:   Vibration Step Stress Failures

Failure Mode	Qty
Broken lead	43
Screws backed out	9
Socket interplay	5
Connector backed out	5
Component fell off (non-soldered)	5
Tolerance issue	4
Card backed out	4
Shorted component	2
Broken component	2
Sheared screws	1
RTV applied incorrectly	1
Potentiometer turned	1
Plastic cracked at stress point	1
Lifted pin	1
Intermittent component	1
Failed component	1
Connectors wearing	1
Connector making intermittent contact	1
Connector broke from board	1
Broken trace	1

Table B.10: **Combined Environment Failures**

Failure Mode	Qty
Broken lead	10
Component fell off (non-soldered)	4
Failed component, cause unknown	3
Broken component	1
Component shorted out	1
Cracked potting material	1
Detached wire	1
Circuit design issue	1
Socket interplay	1

Appendix

C References

Hobbs, Gregg H., *Accelerated Reliability Engineering: HALT and HASS*, John Wiley & Sons Ltd., 2000.

McLean, Harry W., *HALT, HASS & HASA Explained: Accelerated Reliability Techniques*, 2nd Edition, ASQ, 2000.

O'Connor, P.D.T., *Practical Reliability Engineering*, 4th Edition, John Wiley & Sons Ltd., 2001.

Wikipedia, 2010, Definitions for Glossary of Terms.

D Glossary of Terms

Accelerated Life Test (ALT)—The process of determining the useful life of a product in a short period of time by accelerating the use environment (how the product will be used).

Acceleration Factor—Denotes how much you are accelerating the test over the end use conditions.

Activation Energy—The minimum energy required to start a chemical reaction (a chemical reaction is when a substance changes into another substance).

Alpha Risk (α)—Sometimes referred to as a Type I Error, is the producer's risk, or the risk that you *won't* ship a good product because the data indicated that the good sample was bad. In other words, is the risk of rejecting the hypothesis that the mean hadn't changed when it hadn't.

Analysis of Variance (ANOVA)—Estimation of fractional contributions and error variance.

Annualized Failure Rate—The rate at which a product will fail, calculated on a yearly basis.

Arrhenius Model—Describes how component reliability is adversely affected as the temperature increases.

Availability—The probability a system is ready for use when needed or the proportion of time a system is ready for use.

Benchmarking—The process of determining and comparing reliability related metrics for a set of specific products in a specific market. The purpose of Benchmarking is to gain a clear understanding of reliability strengths and weaknesses of competitors' products in your market.

Beta Parameter (β)—The slope of the Weibull Distribution (β).

Beta Risk (β)—Sometimes referred to as a Type II Error, is the consumer's risk, or the risk that you *will* ship a defective product because the data indicated that the failed sample actually is good. In other words, is the risk of accepting the hypothesis that the mean hadn't changed when it had.

Black Box Testing—Functional testing based on requirements with no knowledge of the internal program structure or data.

Boundary Interface Diagram—A diagram used as part of a brainstorming exercise often used during FMEA. It is useful when you have a complex system or if your system has interfaces with other systems. In the diagram, draw lines showing the interfaces between the different systems. The interfaces can either be physical, energy, material, or data.

Burn-In—A form of ESS, running a system at elevated temperature to take advantage of heat as an accelerant. The key variables are the burn-in temperature and the amount of burn-in time.

Certified Reliability Engineer (CRE)—The American Society for Quality has a program to become a Certified Reliability Engineer, CRE. Certification is based on education, experience, and a certification test; periodic recertification is required. The body of knowledge for the test includes: reliability management, design evaluation, product safety, statistical tools, design and development, modeling, reliability testing, collecting and using data, etc.[8]

Closed Loop Corrective Action (CLCA)—To identify, analyze, and correct a problem with a product or process.

Coffin-Manson Model—A model used to describe mechanical fatigue in material and crack growth in solder and other metals due to repeated temperature cycling.

Cold Operating Limit (COL)—During HALT, when stepping the temperature down, the lower temperature limit below which the product ceases to function.

Commercial-Off-The-Shelf (COTS)—Commercially available hardware. This is often used with military and space systems in which military or space grade hardware is too costly or not available or both. Note that COTS does not necessarily mean commercial quality.

---

8. "Certified Reliability Engineer" (2010). Wikipedia. [Online].
   http://en.wikipedia.org/wiki/Certified Reliability Engineer

Competitive Analysis—Process by which you compare your product's reliability performance to competitive products. Use the results as input to your Gap Analysis to determine appropriate next steps to improve your organization's reliability program. The two main types of Competitive Analyses are Competitive HALT and Competitive Teardown Analysis.

Competitive HALT—Subject two or more products to HALT up to their operational/destruct limits and then compare the product margins. With similar technologies, a product that has better margins is generally a more reliable product.

Competitive Teardown Analysis—Compare two or more products by disassembling each and then comparing a number of different attributes. The two key elements to this process are to

1. develop a meaningful set of attributes, and
2. develop an objective scoring system for each attribute.

Component Counterfeiting—Process by which a third-party company tries to copy the design of another company. The counterfeit component could have a whole range of issues from having parameters that don't meet the specifications of the genuine component, all the way to a complete empty package that has no functionality at all.

Component Parameter—Testing When the situation arises in which your supplier can't or won't guarantee component parameters, then you should write a custom specification for the product and test those specific parameters on an ongoing basis, either using a sampling plan or 100% testing.

Component Selection—The purpose of selecting the appropriate components for the particular application and environment is to optimize reliability versus cost in the given use environment. Most designs have a few components that account for the majority of reliability or quality issues, and it is these few components that you should analyze in more detail during the design to mitigate these issues before they occur.

Computational Fluid Dynamics (CFD)—One of the branches of fluid mechanics that uses numerical methods and algorithms to solve and analyze problems that involve fluid flows.[9]

---

9. "Computational Fluid Dynamics" (2010). Wikipedia. [Online].
   http://en.wikipedia.org/wiki/Computational_Fluid_Dynamics

Computer Aided Design (CAD)—The use of computer technology for the process of design and design-documentation.[10]

Conductive Anodic Filament (CAF)—A form of electrochemical metal migration or dendritic growth between two conductors within a printed circuit board (PCB).[11]

Confidence Interval—A particular kind of interval estimate of a population parameter. Instead of estimating the parameter by a single value, an interval likely to include the parameter is given. Thus, confidence intervals are used to indicate the reliability of an estimate. How likely the interval is to contain the parameter is determined by the confidence level or confidence coefficient. Increasing the desired confidence level will widen the confidence interval.[12]

Confidence Level—A confidence interval is always qualified by a particular confidence level, usually expressed as a percentage.[13]

Confidence Limits—The end points of the confidence interval.[14]

Consumer's Risk—See Beta risk (β).

Contract Manufacturer (CM)—A company that specializes in manufacturing products for other companies.

Control Factor—In an experiment, these are all of the factors that you have the ability to change.

Corrective Action—See Closed Loop Corrective Action (CLCA).

Corrective and Preventative Action (CAPA)—See Closed Loop Corrective Action (CLCA).

Corrosion—Deterioration of metals caused by oxidation or chemical action.

---

10. "Computer Aided Design" (2010). Wikipedia. [Online].
    http://en.wikipedia.org/wiki/Computer_Aided_Design
11. "Conductive Anodic Filament" (2010). Wikipedia. [Online].
    http://en.wikipedia.org/wiki/Conductive_Anodic_Filament
12. "Confidence Interval" (2010). Wikipedia. [Online].
    http://en.wikipedia.org/wiki/Confidence_Interval
13. "Confidence Level" (2010). Wikipedia. [Online].
    http://en.wikipedia.org/wiki/Confidence_Level
14. "Confidence Limits" (2010). Wikipedia. [Online].
    http://en.wikipedia.org/wiki/Confidence_Limits

Cosmic Rays—Particles coming from outside the solar system. These particles can interact with electronics and cause temporary upsets in performance.

Creep Degradation—The loss of property or breakdown of a material due to overexposure to heat.

Critical to Quality (CTQ)—Specific, measurable characteristics of a product or process that are necessary for your customers' satisfaction.

Cross-Sectioning—In the context of failure analysis, this is the process of grinding of die and package, usually perpendicular to the surface of the die, to examine defects.

Cumulative Distribution Function (CDF)—Describes the probability that a real-valued random variable X with a given probability distribution will be found at a value less than x. Intuitively, it is the "area so far" function of the probability distribution.[15]

Customer-Specified Goals—When your customers specify the reliability requirements for your product. Mean time between failure (MTBF), mean time to repair (MTTR), availability, dead on arrival (DOA) rate, and return rate are common reliability goals, but there are many others.

Date Code—Product build dates that vendors stamp onto their product. These date codes can be very useful if you discover a failure on the manufacturing line or in the field that may be related to a particular date range of when the product was created.

Dead on Arrival (DOA)—The rate of products that don't work when your customer first receives and installs the product.

Decapsulation—During failure analysis, the process of removing the cover of a semiconductor package.

Defect Insertion Rate—The rate at which you introduce new defects to a software process.

Dendrite Growth—A type of failure that is caused by grain development that resembles the increasingly smaller branches of a tree.

---

15. "Cumulative Distribution Function" (2010). Wikipedia. [Online]. http://en.wikipedia.org/wiki/Cumulative_Distribution_Function

Derating Analysis—The practice of operating at a lower stress condition than the rating specified for a component.

Design FMEA—FMEAs performed on the system at the design level. The purpose is to analyze how failure modes affect the system, and to minimize failure effects upon the system.

Design for Maintainability—Part of the DfX (Design for Excellence) methodology, in which you consider during the design phase the preventive maintenance and repair of your product.

Design for Manufacturability (DFM)—Part of the DfX (Design for Excellence) methodology in which you consider during the design phase how your manufacturing team or partner will assemble and build your product.

Design for Reliability (DFR)—Part of the DfX (Design for Excellence) methodology in which you consider during the design phase how to optimize the reliability of your product.

Design for Warranty—Part of the DfX (Design for Excellence) methodology in which you consider during the design phase how to optimize your warranty period (set the warranty rate for your product to minimize the number of products that fail within your product's warranty period while maximizing the length of warranty you can offer).

Design Margin—See Operating Margin.

Design of Experiments (DOE)—An experiment which can focus on a wide range of key input factors or variables and will determine the optimum levels of each of the factors.

Destruct Limit—In HALT, a limit in which the failure doesn't recover when you reduce or remove the stress.

Detection (D)—In an FMEA, a score that indicates the ability to detect a failure if it does occur.

Detection Shift Level—The shift in percentage from your current production failure rate to a new production failure rate due to a process change. In most cases, you can't detect this shift immediately, resulting in shipping product at this higher failure rate for a period of time until you can detect this shift. The lower the detection shift value, the more samples you must test before you discover this change.

Dielectric Breakdown—The complete failure of a dielectric material that is characterized by a disruptive electrical discharge through the material that is due to deterioration of material or to an excessive sudden increase in applied voltage.

Duty Cycle—The percentage of time a product is operated compared to the total calendar time.

Early Reliability Testing (ERT)—A development strategy that can provide higher reliability and quality, with less cost and time for development, as well as less development risk.

Electrical Overstress (EOS)—An unusual spike in voltage applied to the system that causes damage (usually at the component level), resulting in a failure. Some signs of EOS are blown metal line or molten damage. Electrostatic Discharge (ESD) is a subset of EOS.

Electromagnetic Interference (EMI)—Disturbance that affects an electrical circuit due to either electromagnetic conduction or electromagnetic radiation emitted from an external source. The disturbance may interrupt, obstruct, or otherwise degrade or limit the effective performance of the circuit. The source may be any object, artificial or natural, that carries rapidly changing electrical currents, such as an electrical circuit, the Sun or the Northern Lights.[16]

Electromigration—High current density that can move atoms out of the active regions, leading to emergence of dislocations and point defects, acting as non-radiative recombination centers and producing heat instead of light.[17]

Electron Microscope (EM)—Type of microscope that produces an electronically-magnified image of a specimen for detailed observation. The EM uses a particle beam of electrons to illuminate the specimen and create a magnified image of it. It has a greater resolving power than a light-powered optical microscope, because it uses electrons that have wavelengths about 100,000 times shorter than visible light (photons), and can achieve magnifications of up to 2,000,000x, whereas light microscopes are limited to 2000x magnification.[18]

---

16. "Electromagnetic Interference" (2010). Wikipedia. [Online].
    http://en.wikipedia.org/wiki/Electromagnetic_Interference
17. "Electromigration" (2010). Wikipedia. [Online].
    http://en.wikipedia.org/wiki/Electromigration
18. "Electron Microscope" (2010). Wikipedia. [Online].
    http://en.wikipedia.org/wiki/Electron_Microscope

Electronic Design Automation (EDA)—A category of software tools for designing electronic systems, such as printed circuit boards and integrated circuits.[19]

Electronic Manufacturing Service (EMS)—See Contract Manufacturer (CM).

Electrostatic Discharge (ESD)—The release of static electricity when two objects come into contact. Typically this is when a person has a build-up of charge, comes in contact with electronics, and causes a discharge event, transferring the charge to the electronics.

End-of-Life (EOL)—Analysis An estimate of the onset of wear-out failures for consumables (e.g., electrolytic capacitors, fans, motors, drives). An EOL Analysis can aid in planning warranty costs and scheduling Preventive Maintenance (PM).

Engineering Change Order (ECO)—A system by which you make a change to your product, including all of the documentation involved.

Environmental Stress Screening (ESS)—The process of applying external stresses to a manufacturing sample of a product (usually temperature and/or vibration) to accelerate time in order to expose any manufacturing defects before shipping the sample to a customer. HASS, Run-In, and Burn-In are types of ESS.

Evolutionary Product—A product that is similar to a previous generation but with modifications for new features.

Expert Reliability—Review When a senior engineer (expert) reviews the work of an engineer or junior engineer.

Exponential Distribution—A continuous probability distribution that describes the times between events in a Poisson process, i.e., a process in which events occur continuously and independently at a constant average rate.[20]

Extended Warranty—A warranty contract that you can purchase to increase the amount of time a product is covered under the manufacturer's warranty period.

---

19. "Electronic Design Automation" (2010). Wikipedia. [Online].
    http://en.wikipedia.org/wiki/Electronic_Design_Automation
20. "Exponential Distribution" (2010). Wikipedia. [Online].
    http://en.wikipedia.org/wiki/Exponential_Distribution

Facilitation—The process of designing and running a successful and impartial meeting. Facilitation services the needs of any group who is meeting with a common purpose, whether it be making a decision, solving a problem, or simply exchanging ideas and information.[21]

Facilitation of Code Reliability Review—The use of reliability reviews to target the core and vulnerable sections of code to allow the owner of the source code to develop sufficient synergy with a small team of developers in finding defects.

Facilitation of Team Design Review—Conducting brief, informal reviews that are highly interactive at multiple points throughout the progression, from system architecture all the way to low-level design.

Facilitation of Team Design Template Review—Conducting group pre-design review meetings to provide your team with forums to expand their knowledge base of design techniques by exchanging design templates.

Failure Mechanism—The cause of the failure mode such as "corrosion" or "vibration."

Failure Mode—The actual symptom of the failure, such as "failed component" or "degradation of performance."

Failure Modes and Effects Analysis (FMEA)—A systematic method of identifying and preventing product and process problems *before* they occur.

Failure Modes, Effects, and Criticality Analysis (FMECA)—Similar to an FMEA with the "C" standing for criticality. The process is the same except now you capture one more score—the criticality of the failure. See MIL-HDBK-1629 for a more detailed explanation on the term criticality.

Failure Reporting Analysis and Corrective Action System (FRACAS)—See Closed Loop Corrective Action (CLCA).

Failure Review Board (FRB)—A team assigned to a company to review each failure and determine its disposition.

Fatigue—The effects of dynamic loading causing a loss in strength of a material.

---

21. "Facilitation" (2010). Wikipedia. [Online].
    http://en.wikipedia.org/wiki/Facilitation

Fault Tree Analysis (FTA)—A systematic, deductive method for defining a single specific undesirable event and determining all possible failures that could cause the event in question to occur. A top-down approach to failure mode analysis.

Field Data Tracking—The process of collecting product performance data from the field and analyzing it using statistics to determine trends and to measure your reliability against your original goals.

Field Failure Rate—The rate at which a product fails in the field. This is usually expressed as a percentage.

Finite Element Analysis (FEA)—A technique to estimate the responses of structures and materials to environmental factors such as fluid flow, forces, heat, and vibration. With complex mechanical components, it is possible to model these factors by subdividing a component down into small, "finite" elements, and to analyze the component as an assembly of these small, simple elements.

First Pass Yield—The percentage of samples that pass the testing process the first time through without requiring any rework.

First Year Multiplier (FYM)—A factor applied to a reliability prediction for the increased failure rate due to manufacturing imperfections.

Fourier Transform Infrared Spectroscopy (FTIR)—Technique which is used to obtain an infrared spectrum of absorption, emission, or photoconductivity of a solid, liquid or gas. An FTIR spectrometer simultaneously collects spectral data in a wide spectral range.

Fracture Mechanics—Field of mechanics concerned with the study of the formation of cracks in materials. It uses methods of analytical solid mechanics to calculate the driving force on a crack and those of experimental solid mechanics to characterize the material's resistance to fracture.[22]

Fretting—Wear and sometimes corrosion damage at the edges of contact surfaces. This damage is induced and in the presence of repeated relative surface motion, as induced for example by vibration.[23]

---

22. "Fracture Mechanics" (2010). Wikipedia. [Online].
   http://en.wikipedia.org/wiki/Fracture_Mechanics
23. "Fretting" (2010). Wikipedia. [Online].
   http://en.wikipedia.org/wiki/Fretting

Appendix D: Glossary of Terms

Functional Testing—The method of testing a product to ensure it is performing its intended functions. When performing a reliability test, it is usually important to functionally test the product at all times.

Fundamental Limit of the Technology (FLT)—The limit at which you can't go any higher with a stress without changing the failure mechanism and producing nonrelevant failures.

Gap Analysis—The process of determining the spread between your reliability goals and your current capabilities.

Goal Setting—The process of setting targeted goals at the beginning of a design/development program and then putting forth a plan to achieve the goals.

Golden Nuggets—Those few techniques that your team does well and recognizes they do them well so these become engrained into your culture.

Grms—Gravity Root Mean Squared. A unit of measure for vibration in which you calculate the total amount of the vibration under a Power Spectral Density (PSD) curve within a band of frequencies.

Guard Band Limits—During a HALT, a guideline as to the limits you meet as a minimum.

HALT-to-AFR Calculator—A mathematical model that, when provided with the appropriate HALT and product information, will accurately estimate a product's actual failure rate (AFR) in the field.

Highly Accelerated Life Test (HALT)—A design technique that you can use to discover product weaknesses and design margins. The intent is to subject a product systematically to stress stimuli well beyond the expected field environments in order to determine the operating and destruct limits of your product.

Highly Accelerated Life Test (HALT) Plan—The plan by which you document the information on the types of stresses, levels of stresses, and order of stresses. You should also determine the number of samples, functional tests, what parameters to monitor, and what constitutes a failure. Decisions in this plan will dictate the relative success of the HALT.

Highly Accelerated Life Test (HALT) Report—The report by which you document all of your findings in HALT (including at what point each failure occurred and what steps you took to work around each failure). Pictures, charts, and graphs are great for your management. However, the most important parts of the report are your recommendations and the follow-up actions you plan to take.

Highly Accelerated Stress Audit (HASA)—Similar to HASS, except that you will screen only a portion of the product (called an audit) rather than 100% as in HASS. As with any audit, you will need to set up criteria for when to decrease the sample size and when to increase the sample size, mostly based on the results of the audit.

Highly Accelerated Stress Audit (HASA) Plan—The plan that documents how to move from 100% HASS to a HASS audit, also known as HASA. When writing the plan, make sure to describe what criteria need to be met in order to move to HASA. The two key criteria that you need to satisfy before switching from HASS to HASA are:

1. The defect rate must be at its target, and
2. The process must be stable. In the plan, decide on the following parameters: detection shift level, alpha risk and beta risk levels, and sample size.

Highly Accelerated Stress Screening (HASS)—A process comprising a set of stresses performed on a product before it is shipped with the goal of finding manufacturing related defects. The set of stresses combined together make up the screen.

Highly Accelerated Stress Screen (HASS) Plan—The plan that documents the HASS process from start to finish, including choosing the stress types, developing the screen profile, the equipment trade-off analysis, fixture design, POS, HASS implementation strategy, and the trend analysis. The plan will serve as your roadmap and you can use it as a decision tool during the implementation process.

Hot Carrier Injection—Phenomenon in solid-state or semiconductor electronic devices where either an electron or a "hole" gains sufficient kinetic energy to overcome a potential barrier necessary to break an interface state.[24]

Hot Operating Limit (HOL)—During HALT, when stepping the temperature up, the upper temperature limit above which the product ceases to function.

Human Factors Analysis—The study of all aspects of the way humans relate to the equipment, with the aim of improving operational performance, reliability, and safety.

---

24. "Hot Carrier Injection" (2010). Wikipedia. [Online].
    http://en.wikipedia.org/wiki/Hot_Carrier_Injection

Infant Mortality—The period of a product life cycle in which the failure rate is decreasing in rate over time. Failures in this period of time are typically due to a manufacturing defect.

Infant Mortality Region—The region of decreasing slope or decreasing failure rate from time t=0 and forward. Typically the probability of failure is highest immediately after you ship the product to your customer and then after that point, the failure rate reduces.

Input (in reference to DOE)—These are the entries into the system from another device or from a user (such as a user interfacing with the system).

Internally-Specified Goals—Goals that come from within your company rather than from external competition. These goals are usually based on trying to be better than previous products. One of the executives of a company may put forth some sort of edict such as "our next product will have half the field returns than our previous product." You can then take this and turn it into a goal statement.

Ion Chromatography—Process that allows the separation of ions and polar molecules based on their charge.[25]

Latent Defect Density—The number of defects per lines of code still remaining in the software after you start shipping the software.

Leading Indicator—An ALT that is run not to the point of a failure but rather until you detect an indication that a parameter has changed. First you should determine what parameter(s) you should monitor and then you should run a calibration step in order to be able to properly extrapolate the failure.

Lessons Learned—The process of capturing all of the issues that occur in a program in a centralized system for all personnel to share, and then reviewing these issues prior to embarking on a new design.

Life Cycle Cost (LCC)—Refers to the total cost of ownership over the life of a product. Also commonly referred to as "cradle to grave" or "womb to tomb" costs.[26]

_____

25. "Ion Chromatography" (2010). Wikipedia. [Online].
    http://en.wikipedia.org/wiki/Ion_Chromatography
26. "Life Cycle Cost" (2010). Wikipedia. [Online].
    http://en.wikipedia.org/wiki/Life_Cycle_Cost

Log-Log Graph—A graph drawn on paper with logarithmic horizontal axis and logarithmic vertical axis.

Log-Normal Distribution—A probability distribution of a random variable whose logarithm is normally distributed. A variable might be modeled as log-normal if it can be thought of as the multiplicative product of many independent random variables each of which is positive.

Maintainability Analysis—The method of determining how to best design a product for ease of maintenance.

Maintainability Prediction—The method of determining how long it takes to repair a product once it fails.

Manufacturing Screen—The method of stressing a product either electrically or environmentally to accelerate time during manufacturing with the goal of uncoving latent manufacturing defects that would otherwise escape and then soon fail after your customer starts using the product.

Markov Analysis—Process of analyzing the reliability and availability of systems whose components exhibit strong dependencies.

Mean Square Deviation—Measure of the differences between values predicted by a model or an estimator and the values actually observed from the thing being modeled or estimated.[27]

Mean Time Between Failure (MTBF)—Predicted elapsed time between inherent failures of a system during operation. It is the inverse of the failure rate.[28]

Mean Time to Repair (MTTR)—The average time to repair a product and get it operational after it fails.

Memory Leak—When software routines use up a portion of memory after they load a program and then don't give the memory back after they close the program This causes the available memory to shrink, which in turn results in the product slowing down (because the processor must spend more time looking for available memory).

---

27. "Mean Square Deviation" (2010). Wikipedia. [Online].
    http://en.wikipedia.org/wiki/Mean_Square_Deviation
28. "Mean Time Between Failure" (2010). Wikipedia. [Online].
    http://en.wikipedia.org/wiki/Mean_Time_Between_Failure

MIL217G—An effort being developed by the VITA 51 Working Group to investigate the state of the Reliability Prediction industry and develop a method to address electronics failure rate prediction issues. This standard is meant to replace MIL-HDBK-217.

MIL-HDBK-217—In the 1950s, the Department of Defense (DOD) first standardized electronics Reliability Predictions through the analysis of historical data. This led to the publication of the first edition of MIL-HDBK-217 in 1961, providing the basis of Reliability Predictions that is still widely used today.

Monte Carlo Simulation—A class of computational algorithms that rely on repeated random sampling to compute their results. Because of their reliance on repeated computation of random or pseudo-random numbers, these methods are most suited to calculation by a computer and tend to be used when it is unfeasible or impossible to compute an exact result with a deterministic algorithm.[29]

No Problem Found (NPF)—When your product fails in your customer's hands and you can't duplicate this failure when you retest in your facility.

Noise Factor—Effect of all the uncontrollable factors in an experiment.

Nonlinear Finite Element Analysis (FEA)—A form of Finite Element Analysis which is more general and includes the full spectrum of effects such as large displacement, contact, large strain, and nonlinear material response.

Non-Recurring Engineering (NRE)—Engineering resources that you use once or for a period of time in a program and then stop.

Nonrelevant Failure—Failures modes that occur during reliability testing which would not occur in the field during normal operation.

Normal Distribution—A continuous probability distribution that often gives a good description of data that cluster around the mean. The graph of the associated probability density function is "bell"-shaped, with a peak at the mean, and is known as the Gaussian function or bell curve.[30]

---

29. "Monte Carlo Simulation" (2010). Wikipedia. [Online].
    http://en.wikipedia.org/wiki/Monte_Carlo_Simulation
30. "Normal Distribution" (2010). Wikipedia. [Online].
    http://en.wikipedia.org/wiki/Normal_Distribution

Obsolescence—The process of going through your parts lists to determine which vendors will be discontinuing their parts and when so that you can mitigate this before the event takes place.

Ongoing Reliability Test (ORT)—A method of deriving a reliability figure through testing, typically during the manufacturing process.

Ongoing Reliability Test (ORT) Decision Matrix—During an ORT, there are many different parameters for the test as well as choices for each parameter. You can enter all of these into a matrix to help decide on the optimal combination of values. These then get entered into the ORT Plan.

Ongoing Reliability Test (ORT) Plan—Document the different parameter values you have chosen from the ORT Decision Matrix, including types of stresses, number of samples, length of test, and confidence, along with advantages and disadvantages for each.

Operating Limit—In HALT, a limit in which the failure recovers when you reduce or remove the stress.

Operating Margin—The difference between the operating limit and the product specification.

Optical Microscopy—Uses visible light and a system of lenses to magnify images of small samples.

Original Design Manufacturer (ODM)—A company which designs and manufactures a product which is specified and eventually branded by another firm for sale. Such companies allow the brand firm to produce (either as a supplement or solely) without having to engage in the organization or running of a factory.[31]

Original Equipment Manufacturer (OEM)—Manufacturer of the products or components that are purchased by a company and retailed under the purchasing company's brand name.[32]

---

31. "Original Design Manufacturer" (2010). Wikipedia. [Online].
    http://en.wikipedia.org/wiki/Original_Design_Manufacturer
32. "Original Equipment Manufacturer" (2010). Wikipedia. [Online].
    http://bit.ly/O_E_M
    en.wikipedia.org/wiki/Original_Equipment_Manufacturer

Orthogonal Array—In a DOE, this type of array enables a fair comparison of tolerance factor main effects and advanced statistical tools, like Analysis of Variance (ANOVA). It also enables an estimation of fractional contributions and error variance.

Out of Box Audit (OOBA)—The process of randomly taking a boxed-up system from the shipping area, opening it up, and performing an inspection and/or functional test on the sample in order to measure outgoing quality and reliability.

Output (in reference to DOE)—What the system is supposed to do or the results the system is supposed to produce.

Parameter Diagram (P-Diagram)—A diagram to help focus the brainstorming into four different areas: Piece to Piece Variations, Environment, Customer Usage/Duty Cycle, and Deterioration. Each area is called a Noise Factor, or a factor that you can't control. Start with one area and brainstorm all of the failure modes in this area and then move on to the next area. For each failure mode, you also have Inputs, Outputs, Control Factors, and Error States.

Pareto Chart—A type of chart that contains both bars and a line graph, where individual values are represented in descending order by bars, and the cumulative total is represented by the line. The chart was named after Vilfredo Pareto.[33]

Peck's Model—A model which predicts the acceleration factor of a test based on the stresses of temperature and humidity.

Periodic HALT—HALT that is performed on a scheduled interval in order to discover issues that may enter into the design either from design changes, vendor changes, or vendor process changes.

Phase Transition Point—The transformation of a material from one phase of matter to another.

Physics of Failure (POF)—The process of using knowledge of root-cause failure processes to prevent product failures through product design and manufacturing practices.

Physics of Failure (POF) Model—A model which predicts the behavior of materials, including when and how they will fail.

---

33. "Pareto Chart" (2010). Wikipedia. [Online].
    http://en.wikipedia.org/wiki/Pareto_Chart

Pick and Place Equipment—Equipment used to install components onto a circuit board.

Power Spectral Density (PSD)—Describes how the power of a signal or time series is distributed with frequency.[34]

Precipitation/Detection Screen—This is a HASS technique that consists of a two part profile. In the first part, the precipitation portion, develop your profile to actually go beyond your operating limit but within the destruct limit. As your product approaches its operating limit, make sure to shut it off. Then, turn it back on when it comes back within the operating limit. In the second part, the detection portion, keep your profile within the operating limits. Note that each portion of the screen can be one or more cycles and each portion does not have to be the same number of cycles.

Prediction to ORT Factor—A Reliability Prediction may not be able to give an exact MTBF number, but it will give a number close enough to help determine an ORT sample size and acceleration factor necessary to provide the necessary data on-going. After collecting sufficient data in the ORT, compare your ORT results with your prediction so that you have this difference. You can then use this difference to develop a factor that you can use for future predictions.

Prediction-to-Field Factor—The ratio between the prediction results from previous predictions with the field results after deploying the product. After collecting sufficient data from the field, compare your field results with your prediction so that you have this difference. You can then use this difference to develop a factor that you can use for future predictions.

Preventive Maintenance (PM)—A procedure of inspecting, testing, and reconditioning a system at regular intervals, usually because of a component(s) that wears over time and will cause a predictable failure.

Printed Circuit Board (PCB)—Used to mechanically support and electrically connect electronic components using conductive pathways, tracks or signal traces etched from copper sheets laminated onto a non-conductive substrate. It is also referred to as printed wiring board (PWB). A PCB populated with electronic components is a printed circuit assembly (PCA), also known as a printed circuit board assembly (PCBA).[35]

---

34. "Power Spectral Density" (2010). Wikipedia. [Online].
    http://en.wikipedia.org/wiki/Power_Spectral_Density
35. "Printed Circuit Board" (2010). Wikipedia. [Online].
    http://en.wikipedia.org/wiki/Printed_Circuit_Board

PRISM®—Ties together several tools into a comprehensive system Reliability Prediction methodology. This concept accounts for the myriad of factors that can influence system reliability, combining all those factors into an integrated system Reliability Assessment resource.

Probabilistic Evaluation—When randomness is present, and variable states are not described by unique values, but rather by probability distributions.[36]

Probability of Occurrence (P)—In an FMEA, a score that indicates how often the failure will occur.

Probability Ratio Sequential Test (PRST)—Based on the ratio of an acceptable MTBF (which should have a high probability of acceptance) to an unacceptable MTBF (which should have a low probability of acceptance). They are set up as either testing for a pre-determined amount of time or until a pre-determined number of failures occur.

Process Benchmarking—Comparison between two or more products to determine what techniques each team uses in developing their product. It entails comparing process methodologies such as in-house versus outsource builds, quality philosophy, and screening methods.

Process Capability—The inherent failure rate of a particular process.

Process FMEA—FMEAs performed on the manufacturing processes. They are conducted through the quality planning phase as an aid during production. The purpose is to analyze and correct the possible failure modes in the manufacturing process, including limitations in equipment, tooling, gauges, operator training, or potential sources of error.

Producer's Risk—See Alpha Risk (α)

Product Benchmarking—Comparing product reliability related metrics for a set of specific products in a specific market such as mean time between failure (MTBF), annualized failure rate (AFR), and dead on arrival (DOA) rate.

Product Life Cycle (PLC)—The life of a product in the market with respect to business/commercial costs and sales measures.

---

36. "Probabilistic Evaluation" (2010). Wikipedia. [Online]. http://en.wikipedia.org/wiki/Probabilistic_Evaluation

Product Lifecycle Management (PLM) Tool—A tool that aids companies in managing the entire lifecycle of a product efficiently and cost-effectively, from concept, into design and manufacturing, through service and disposal.

Product Warranty—A warranty contract that comes with your product purchase and covers you for a period of time after you buy the product.

Prognostics—Study of being able to predict failures before they happen.

Proof of HASS Strength—Method by which you can determine if the screen you chose for HASS is strong enough to find defects. In other words, this method proves the screen is effective.

Proof of Screen (POS)—Process of ensuring that the screen you have developed is both safe and effective. The POS consists of two different components:

1. Safety of Screen (to prove the screen is safe), and
2. Proof of HASS Strength (to prove the screen is effective).

Rayleigh Distribution—A continuous probability distribution in which the components are uncorrelated and normally distributed with equal variance. The distribution is named after Lord Rayleigh.[37]

Re-HALT—The process of performing HALT later in the development process after the product has matured, when more samples are available, and test routines are more complete.

Relevant Failure—Failures modes that occur during reliability testing which would occur in the field during normal operation.

Reliability Allocation—See Reliability Apportionment.

Reliability Apportionment—Take your reliability goal and budget portions of it to each of the different assemblies in your system.

Reliability Assessment—See Reliability Program Assessment.

Reliability Bathtub Curve—A relationship between failure rate versus time that expresses the three distinct phases of a product life cycle—infant mortality, steady state, and wear-out.

---

37. "Rayleigh Distribution" (2010). Wikipedia. [Online].
   http://en.wikipedia.org/wiki/Rayleigh_Distribution

Reliability Block Diagram—Diagram often used during Reliability Apportionment to show the contribution of different portions of your system to the overall system reliability.

Reliability Case—Special form of Reliability Program Plan in which the supplier guarantees that their product will meet an agreed set of in-service reliability requirements. The onus of responsibility is on the supplier to build the case by gathering evidence showing that the product will meet the reliability requirements. The supplier then develops a Reliability Case Report, which contains a summary of the Reliability Case with supporting evidence.

Reliability Critical Item List—List of components that require special attention and reduction of this list is a key goal early in a program. Some of the reasons for putting items on this list are: low reliability, high criticality, and long-lead time.

Reliability Demonstration Test (RDT)—The process of demonstrating the steady state reliability of a product through testing.

Reliability Demonstration Test (RDT) Decision Matrix—During an RDT, there are many different parameters for the test as well as choices for each parameter. You can enter all of these into a matrix to help decide on the optimal combination of values. These then get entered into the RDT Plan.

Reliability Demonstration Test (RDT) Plan—Document the different parameter values you have chosen from the RDT Decision Matrix, including types of stresses, number of samples, length of test, and confidence level, along with advantages and disadvantages for each.

Reliability Goal—A statement you make early in the reliability program and enter into your Reliability Program Plan that defines the reliability you plan on achieving for the product. The goal consists of the following four elements:

1. Probability of product performance,
2. Intended function,
3. Specified life, and
4. Specified operating conditions.

Reliability Integration—The process of seamlessly, cohesively integrating reliability techniques together to maximize reliability and at the lowest possible cost.

Reliability Maturity Matrix—Method of categorizing responses and coming up with a summary of where your organization is compared to the rest of the industry. This tool is used to map the results of the Reliability Program Assessment.

Reliability Metrics—Provide the measurements and milestones, the "are we there, yet?" feedback that your organization needs to ensure you are on track toward meeting your goals.

Reliability Performance Reporting—A method of reporting back how you are doing against your Reliability Program Plan.

Reliability Prediction—A method of calculating the reliability of a product or piece of a product from the bottom up by assigning a failure rate to each individual component and then summing all of the failure rates.

Reliability Program Assessment—A detailed evaluation of your organization's approach and processes involved in creating reliable products. The assessment captures the current state of your organization and leads to an actionable Reliability Program Plan (RPP). The results of the assessment are mapped onto the Reliability Maturity Matrix.

Reliability Program Integration Plan (RPIP)—Similar to an RPP but with a special emphasis on ensuring all of the reliability techniques you choose as part of your RPP are integrated together to ensure all areas are covered without duplicating efforts.

Reliability Program Plan (RPP)—A plan that ties together customer requirements, business opportunities and employee opportunities. It includes a reliability goal statement, along with supporting evidence and methodologies on how you plan to achieve this reliability goal.

Reliability Test Goal—Similar to a Reliability Goal but set during a reliability test plan to set a goal for the reliability testing.

Reliability Test Plan (RTP)—The high level plan that calls out all of the reliability testing that you will perform on a product.

Repair Depot—Organization responsible for repairing products and tracking the repair actions.

Return on Investment (ROI)—Analysis which compares the amount spent on a process compared to the amount saved.

Return Rate—The percent of product returned in a given period of time.

Revolutionary Product—A product unlike any product in the past with a brand new application or technology.

Risk Management—The process of discovering, evaluating, and mitigating risks in your product.

Risk Priority Number (RPN)—A measurement system to determine the amount of risk for a given failure mode. An RPN is the multiplication of three components—Severity of Failure (S), Probability of occurrence (P), and Detection (D).

Root Cause Analysis (RCA)—The investigative process to determine the underlying event(s) responsible for a failure.

Run-In—A form of ESS, powering on a system and running tests without the use of accelerated environmental stresses. The key variable is the run-in time.

Safety of Screen—Demonstration that the chosen screen leaves samples with sufficient life left in them to survive a normal lifetime of field use. In other words, this method proves that a screen is safe.

Seeded Samples—A sample of the product that you purposely insert defects to ensure that the screen is able to detect these types of defects.

Service-Affecting Reliability—Rather than considering every component failure, this reliability figure discounts failures which don't affect system performance.

Serviceability—The ability of technical support personnel to install, configure, and monitor computer products, identify exceptions or faults, debug or isolate faults to root cause analysis, and provide hardware or software maintenance in pursuit of solving a problem and restoring the product into service.[38]

Severity of Failure (S)—In an FMEA, a score that indicates how severe the effects will be if the failure mode does occur.

---

38. "Serviceability" (2010). Wikipedia. [Online].
    http://en.wikipedia.org/wiki/Serviceability

Signal-to-Noise (S/N)—A measure used in science and engineering to quantify how much a signal has been corrupted by noise. It is defined as the ratio of signal power to the noise power corrupting the signal. A ratio higher than 1:1 indicates more signal than noise.[39]

Simulation Program with Integrated Circuit Emphasis (SPICE)—Program that simulates electronic circuits.

Software Bug Tracking Database—Central repository for all software errors found during the software product life cycle. The databased also contains the status of each bug as well as the final disposition.

Software Design for Reliability (SDFR)—Similar methodology as Design for Reliability except with Software.

Software Failure Modes and Effects Analysis (SFMEA)—FMEAs that focus on potential software bugs as well as errors in interfaces and errors in boundary conditions. This is an excellent technique if you have a set of bugs and are trying to determine the likely cause.

Software Fault Tolerance—Software that is designed to keep a system working to a level of satisfaction in the presence of faults.

Software Phase Containment Metric Tracking—A method of tracking software bugs to prevent a particular class of bug from reappearing. This is similar to the corrective action portion of the Hardware RCA where you not only need to fix the problem and prevent that particular problem from recurring, you need to also fix the process that caused the problem in order to show continual improvement.

Software Prediction Model—A model which provides estimates of the number of faults in the resulting software; greater consistency in reliability leads to increased accuracy in the modeling output.

Software Reliability Assessment—Similar to a Reliability Program Assessment but specifically focusing on the software team.

Software Robustness and Coverage Testing Techniques—Techniques effective at finding software failures during software testing such as fault injection.

---

39. "Signal-to-Noise" (2010). Wikipedia. [Online].
    http://en.wikipedia.org/wiki/Signal-to-Noise

Steady State—The period of a product life cycle in which the failure rate is constant over time, or independent of the amount of time that has passed.

Stress Analysis—Engineering discipline that determines the stress in materials and structures subjected to static or dynamic forces or loads.[40]

Structural Fatigue—Progressive and localized structural damage that occurs when a material is subjected to cyclic loading. The nominal maximum stress values are less than the ultimate tensile stress limit, and may be below the yield stress limit of the material.[41]

Supportability—Deals with all the aspects related to the maintenance, repair and support of systems and products to ensure continued operation or functioning of the systems or product(s).[42]

Technology Risk Assessment—The identification, categorization and prioritization of hardware and software risks to achieve key reliability business objectives.

Telcordia SR1171—Titled "Methods and Procedures for System Reliability Analysis," this special report issued by Telcordia offers guidelines around different types of modeling methods.

Telcordia SR332—Titled "Reliability Prediction Procedure for Electronic Equipment," this special report issued by Telcordia offers guidelines around performing a reliability prediction on a product. As part of the report, there is a table of failure rates for many common components in use today.

Temperature Cycling—A form of ESS, cycling the temperature of a system between cold and hot. The key variables here are temperature ranges, temperature rate of change, dwell time at each temperature extreme, and the number of cycles.

Test/Analyze/Fix—A form of testing program in which you test a product to failure, analyze the failure, fix the failure, and test the product again but apply more stress. This is an iterative process in order to make the product more robust.

---

40. "Stress Analysis" (2010). Wikipedia. [Online].
    http://en.wikipedia.org/wiki/Stress_Analysis
41. "Structural Fatigue" (2010). Wikipedia. [Online].
    http://en.wikipedia.org/wiki/Structural_Fatigue
42. "Supportability" (2010). Wikipedia. [Online].
    http://en.wikipedia.org/wiki/Supportability

Thermal Analysis—To estimate the temperature distribution throughout a product based on the thermal boundary conditions and specified heat sources.

Thermo-Gravimetric Analysis (TGA)—A type of testing that is performed on samples to determine changes in weight in relation to change in temperature. Such analysis relies on a high degree of precision in three measurements: weight, temperature, and temperature change.[43]

Thermo-Mechanical Analysis (TMA)—The measurement of a change of a dimension or a mechanical property of the sample while it is subjected to a temperature regime.

Time Dependent Dielectric Breakdown—A failure mechanism in MOSFETs, when the gate oxide breaks down as a result of long-time application of relatively low electric field (as opposite to immediate breakdown, which is caused by strong electric field). The breakdown is caused by formation of a conducting path through the gate oxide to substrate due to electron tunneling current, when MOSFETs are operated close to or beyond their specified operating voltages.

Tolerance Design—The use of Design of Experiments (DOE) techniques, along with economic considerations, to control the output variation of a design.

Tolerance Stack Up—The use of Design of Experiments (DOE) techniques, along with economic considerations, to control the output variation of a design.

Trade-Off Analysis—Method of performing several different analyses with the purpose of comparing the results of each to determine the best course of action.

Type I Error—See Alpha Risk.

Type II Error—See Beta Risk

Uprating—Using a component outside of its specifications and determining how much of a reliability impact this will have on your design. This is the opposite of derating.

User FMEA—FMEAs that focus specifically on the end user and how they will use, misuse, or possibly even abuse your product. An input to the User FMEA is the user manual. The User FMEA will look at installation, use, and end-of-life

---

43. "Thermo-Gravimetric Analysis" (2010). Wikipedia. [Online].
   http://en.wikipedia.org/wiki/Thermo-Gravimetric_Analysis

situations. Whenever a user is involved, you should pay specific attention to the possibility of the user using the product incorrectly, risking either the integrity of the product, or worse, creating an unsafe situation.

Vendor Assessment—A systematic evaluation of a broad range of potential reliability activities and techniques as currently employed and integrated with one or more vendors.

Verification HALT—After you perform HALT and provide corrective action for a specific failure, then perform HALT again to ensure that the corrective action improved the product performance and did not introduce new failure modes. This second HALT is called a Verification HALT.

Vibration Operating Limit (VOL)—During HALT, when stepping the vibration up, the limit above which the product ceases to function.

VITA 51 Working Group—Formed in 2004 to investigate the state of the Reliability Prediction industry and develop a method to address electronics failure rate prediction issues. They found that the MIL-HDBK-217 method had become obsolete compared with current electronics technologies; however, it remained the most common method used in industry to predict electronics reliability.

Warranty—A guarantee given to the purchaser by a company stating that a product is reliable and free from known defects and that the seller will, without charge, repair or replace defective parts within a given time limit and under certain conditions.

Warranty Analysis—Part of the Design for Warranty (DFW) methodology in which you use warranty goals, strategies, and data to jump-start your analysis with the development team during product design. The objective is to identify and prioritize the appropriate warranty metrics, goals, strategies, and action plans to reduce warranty expenses.

Warranty Burden—This defines who is responsible for warranty claims if a product fails within the warranty period. The warranty burden isn't uniformly distributed throughout the design and manufacturing supply chain. In the supply chain, you have contract manufacturers (CM's), original design manufacturers (ODM's), and component suppliers. Companies are starting to ask their supply chain to share the burden of warranty costs.

Warranty Cost Analysis—Method of predicting the amount of money a company will have to pay during the warranty period of a product.

Warranty Event—When a product fails within the warranty period.

Warranty Expenditure—The amount spent on warranty events.

Warranty Period—Period of time that a product is covered under a warranty.

Warranty Projection—A prediction for the number of products (or amount of money) that will be spent on products failing within the warranty period.

Warranty Reserve—The money set aside to cover the cost of products failing within the warranty period.

Warranty Review—The identification and prioritization of warranty performance tracking and cost reduction opportunities. This is essentially the metric check for the Warranty Analysis that you performed in the design phase.

Wear-Out—The period of a product life cycle in which the failure rate is increasing in rate over time. Failures in this period of time are typically due to end-of-life events.

Wear-Out Mechanism—A failure mechanism that is related to an end-of-life event, such as a breakdown of a material.

Weibull Distribution—A continuous probability distribution that can take many different shapes depending on the value of the shape parameter (β). A $\beta < 1$ exhibits a failure rate that decreases with time. A $\beta = 1$ exhibits a failure rate that is constant with respect to time. A $\beta > 1$ exhibits a failure rate that increases with time. All three different phases of the Reliability "Bathtub" Curve can be modeled with the Weibull distribution and varying values of. The distribution was named after Waloddi Weibull who described it in detail in 1951.[44]

Workaround—Situations where you have a failure and apply a temporary change or fix so that you can continue with the testing, knowing that you will have to go back after the completion of the test and perform a more detailed failure analysis and provide a more permanent fix.

X-Ray Fluorescence (XRF)—The emission of characteristic "secondary" (or fluorescent) x-rays from a material that has been excited by bombarding with high energy x-rays or gamma rays. The phenomenon is widely used for chemistry analysis, particularly in the investigation of metals, glass, and ceramics.[45]

---

44. "Weibull Distribution" (2010). Wikipedia. [Online].
    http://en.wikipedia.org/wiki/ Weibull_Distribution
45. "X-Ray Fluorescence" (2010). Wikipedia. [Online].
    http://en.wikipedia.org/wiki/X-Ray_Fluorescence

E   Acronyms

°C	Degrees Centigrade
AFR	Annualized Failure Rate or Average Failure Rate
ALT	Accelerated Life Testing
ANOVA	Analysis of Variance
ASME	American Society of Mechanical Engineers
ASP	Authorized Service Provider
ASQ	American Society of Quality
AVL	Approved Vendor List
CAD	Computer Aided Design
CAF	Conductive Anodic Filament
CALCE	Center for Advanced Life Cycle Engineering (part of University of Maryland)
CAPA	Corrective and Preventive Action
CDF	Cumulative Distribution Function
CFD	Computational Fluid Dynamics
CLCA	Closed Loop Corrective Action
CM	Contract Manufacturer

COL	Cold Operating Limit (see also LOL)
COTS	Commercial-off-the-Shelf
CPK	Process Capability
CRE	Certified Reliability Engineer
CRM	Customer Relational Management
CTQ	Critical to Quality
DFM	Design for Manufacturability
DFR	Design for Reliability
DFW	Design for Warranty
DOA	Dead on Arrival
DOD	Department of Defense
DOE	Design of Experiments
DTIC	Defense Technical Information Center
DVT	Design Verification Test
ECO	Engineering Change Order
EDA	Electronic Design Automation
EMI	Electromagnetic Interference
EMS	Electronic Manufacturing Service
EOL	End-of-Life
EOS	Electrical Overstress
ERT	Early Reliability Testing
ESD	Electrostatic Discharge
ESR	Equivalent Series Resistance
FEA	Finite Element Analysis

FET	Field Effect Transistor
FLT	Fundamental Limit of Technology
FMEA	Failure Modes and Effects Analysis
FMECA	Failure Modes Effects and Criticality Analysis
FRACAS	Failure Review Analysis and Corrective Action System
FRB	Failure Review Board
FTA	Fault Tree Analysis
FTIR	Fourier Transform Infrared
FYM	First Year Multiplier
GRMS	Gravity Root Mean Squared
HALT	Highly Accelerated Life Test
HASA	Highly Accelerated Stress Audit
HASS	Highly Accelerated Stress Screen
HDD	Hard Disk Drive
HOL	High Operating Limit (also see UOL)
IC	Integrated Circuit
IT	Information Technology
IEEE	Institute of Electrical & Electronics Engineers
IP	Intellectual Property
KHZ	Kilohertz
KSLOC	Kilo Source Lines of Code
LCC	Life Cycle Cost
LDL	Lower Destruct Limit
LMM	Lumped Mass Model

LOL	Lower Operating Limit (see also COL)
MSD	Mean Square Deviation
MTBF	Mean Time Between Failure
MTTR	Mean Time to Repair
NPF	No Problem Found
NRE	Nonrecurring Engineer
NSF	National Science Foundation
ODM	Original Design Manufacturer
OEM	Original Equipment Manufacturer
OOBA	Out of Box Audit
ORT	Ongoing Reliability Test
PATCA	Professional and Technical Consultants Association
PCB	Printed Circuit Board
PHM	Prognostic and Health Management
PLC	Product Life Cycle
PLM	Product Lifecycle Management
PM	Preventive Maintenance
POF	Physics of Failure
POS	Proof of Screen
PRG	Product Realization Group
PRN	Product Realization Network
PRST	Probability Ratio Sequential Testing
PTH	Plated Through Hole
RCA	Root Cause Analysis

RDT	Reliability Demonstration Test
RiAC	Reliability Information Analysis Center
ROI	Return on Investment
RPIP	Reliability Program and Integration Plan
RPN	Risk Priority Number
RPP	Reliability Program Plan
RTP	Reliability Test Plan
S/N	Signal-to-Noise
S-N	Stress versus Number of Cycles Relationship
SDFR	Software Design for Reliability
SFMEA	Software Failure Modes and Effects Analysis
SFTA	Software Fault Tree Analysis
SME	Society of Manufacturing Engineer
SPICE	Simulation Program with Integrated Circuit Emphasis
SRC	System Reliability Center
TGA	Thermo-Gravimetric Analysis
TMA	Thermo-Mechanical Analysis
TRIAC	Triode for Alternating Current
UDL	Upper Destruct Limit
USB	Universal Serial Bus
UOL	Upper Operating Limit (also see HOL)
VOL	Vibration Operating Limit
VDL	Vibration Destruct Limit
XRF	X-Ray Fluorescence

About the Author

Mike Silverman

Mike is founder and managing partner at Ops A La Carte LLC®, a professional consulting company that has an intense focus on helping clients with reliability throughout their product life cycle. Mike has over 25 years' experience in reliability engineering, reliability management, and reliability training. He is an experienced leader in reliability improvement through analysis and testing. Through Ops A La Carte®, Mike has had extensive experience as a consultant to high-tech companies. A few of the main industries are aerospace and defense, clean technology, consumer electronics, medical, networking, oil and gas, semiconductor equipment, and telecommunications. Most of the examples in this book have been taken from Mike's experiences.

Mike is an expert in accelerated reliability techniques and owns HALT and HASS Labs®, one of the oldest and most experienced reliability labs in the world. Mike has authored and published 25 papers on reliability techniques and has presented these in countries around the world, including Canada, China, Germany, Japan, Korea, Singapore, Taiwan, and the USA. He has also developed and currently teaches over 30 courses on reliability techniques. Mike has a BS degree in electrical and computer engineering from the University of Colorado at Boulder, and is a Certified Reliability Engineer (CRE) through the American Society for Quality (ASQ). Mike is a member of the American Society for Quality (ASQ), Institute of Electrical and Electronics Engineers (IEEE), Product Realization Group (PRG), Professional and Technical Consultants Association (PATCA), and IEEE Consulting Society. Mike is currently the IEEE Reliability Society Santa Clara Valley Chapter Chair.

You can contact Mike via the Ops A La Carte® website at http://www.opsalacarte.com.

Index

W

X

Other Happy About® Books

Purchase these books at Happy About http://happyabout.info or at other online and physical bookstores.

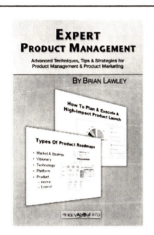

Expert Product Management

This book teaches both new and seasoned Product Managers and Product Marketers powerful and effective ways to ensure they give their products the best possible chance for success.

Paperback $19.95
eBook $14.95

42 Rules of Product Management

42 Rules of Product Management is a collection of product management wisdom from forty experts from around the world. This book will help you lead with greater effectiveness and influence.

Paperback $19.00
eBook $14.00

Agile Excellence for Product Managers

Agile Excellence for Product Managers is a plain-speaking guide on how to work with Agile development teams to achieve phenomenal product success.

Paperback $24.95
eBook $14.95

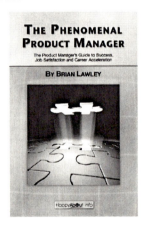

The Phenomenal Product Manager

This book goes beyond the basics and teaches you how to work more effectively with your teams, how to influence when you have no formal authority, how to get the most important work done in less time and how to manage and accelerate your career.

Paperback $19.00
eBook $14.00

A Message from Happy About®

Thank you for your purchase of this Happy About book. It is available online at http://happyabout.com/productreliability.php or at other online and physical bookstores.

- Please contact us for quantity discounts at sales@happyabout.info
- If you want to be informed by email of upcoming Happy About® books, please email bookupdate@happyabout.info

Happy About is interested in you if you are an author who would like to submit a non-fiction book proposal or a corporation that would like to have a book written for you. Please contact us by email editorial@happyabout.info or phone (1-408-257-3000).

Other Happy About books available include:

- #LEADERSHIPtweet Book01:
  http://www.happyabout.com/thinkaha/leadershiptweet01.php
- #PARTNER tweet Book01:
  http://www.happyabout.com/thinkaha/partnertweet01.php
- Expert Product Management:
  http://www.happyabout.com/expertproductmanagement.php
- 42 Rules of Product Management:
  http://www.happyabout.com/42rules/42rulesproductmanagement.php
- 42 Rules to Increase Sales Effectiveness:
  http://www.happyabout.com/42rules/increasesaleseffectiveness.php
- 42 Rules for Driving Success With Books:
  http://www.happyabout.com/42rules/books-drive-success.php
- Agile Excellence for Product Managers:
  http://www.happyabout.com/agileproductmangers.php
- The Phenomenal Product Manager:
  http://www.happyabout.com/phenomenal-product-manager.php
- #QUALITYtweet Book01:
  http://www.happyabout.com/thinkaha/qualitytweet01.php
- Scrappy Project Management®:
  http://www.happyabout.com/scrappyabout/project-management.php
- #PROJECT MANAGEMENT tweet Book01:
  http://www.happyabout.com/thinkaha/projectmanagementtweet01.php

CPSIA information can be obtained at www.ICGtesting.com
Printed in the USA
LVOW11*1823250816

501855LV00009B/103/P